Klaus Joachim Zülch • Otfrid Foerster, Physician and Naturalist

Otfrid Foerster · Physician and Naturalist

November 9, 1873 – June 15, 1941

By Klaus Joachim Zülch

Translated from the German by
Adolf Rosenauer and Joseph P. Evans

Springer-Verlag Berlin · Heidelberg · New York 1969

Klaus Joachim Zülch, M. D.
apl. Professor of Neurology and Psychiatry, University Köln
Director of the Max-Planck-Institut for Brain Research, Department of General Neurology, and the Neurological Clinic of the City Hospital, Köln-Merheim

Translators:

Adolf Rosenauer, M. D.
Neurosurgery, Neurology, Electroencephalography
890 Mill Street, Reno, NV 89502/USA

J. P. Evans, M. D.
Professor of Neurological Surgery, University of Chicago, IL 60637/USA

ISBN 978-3-642-49520-5 ISBN 978-3-642-49809-1 (eBook)
DOI 10.1007/ 978-3-642-49809-1

German Edition Published 1966 with the Title "Otfrid Foerster · Arzt und Naturforscher"
Springer-Verlag Berlin · Heidelberg · New York

Softcover reprint of the hardcover 1st edition 1969

Library of Congress Catalog Card Number 75–95563

Title-No. 1619

Translators' Note

The appearance in 1966 of K. J. ZÜLCH's tribute to the German neurologist and neurological surgeon, OTFRID FOERSTER, marked the twenty-fifth anniversary of the death of this distinguished physician and neuroscientist to whom the neurology of the English speaking world owes so much. His influence during his life time was perhaps greater on the North American continent than in other English speaking countries, though much of his inspiration derived from the work of the Englishman, J. HUGHLINGS JACKSON.

The translation into English of Doctor ZÜLCH's summation of FOERSTER's life and work seemed to us important because of its intrinsic value, but no less important is the illustration of FOERSTER's method of analysis of central nervous system function.

One of us (J. P. E.) had the privilege of spending some months in Breslau in 1936 working at the Neurologisches Forschungsinstitut, experiencing the warmth of FOERSTER's personality, and sharing in the hospitality of the FOERSTER's home. Thus, this translation offers in a sense an opportunity to repay, in part, a debt to "the Professor", as he was affectionately known to his juniors—an affection which, interestingly, was probably greater among his Anglo-Saxon students than among his own people. This reflected his devotion to English and American neurology and neurological surgery.

The initial and literal translation has been carried out by the younger of the team (A. R.), a graduate of the Medical Faculty of Innsbruck University and now a practicing neurological surgeon in the United States. The accuracy of the translation is due to him. Any failure to render the translation into the American idiom is the responsibility of J. P. E.

Reno and Chicago, August 1969 — The Translators

Introduction

The 15th of June, 1966 marked the 25th anniversary of the death of OTFRID FOERSTER, one of those scientists who achieve international recognition in their younger years. He ranks among the greatest of the German neurologists and he stands as a peer among the great names of the world responsible for the shaping of the image of Neurology: HUGHLINGS JACKSON, CHARCOT, DUCHENNE DE BOULOGNE, DEJERINE, ERB, Sir HENRY HEAD, MONAKOW, and SHERRINGTON.

As if possessed by a demon FOERSTER spent himself indefatigably to achieve his almost superhuman task during his lifetime. His work will remain a determinant for Neurology for a long time to come.

Let his own publications speak for him. The epilogues, written shortly after his death, picture him more vividly than can be done today. A short biography and a number of memoirs introduce excerpts from his scientific presentations, many of which are introduced by a brief prefatory note regarding the circumstances of the specific project.

Our gratitude is extended to the Springer Publishing House for facilitating an English translation of the original German edition prepared for the Joint Annual Meeting of the "German Society for Neurology" and the "German Society for Internal Medicine", 1966. And I should also like to express my gratitude to Dr. ADOLF ROSENAUER and Dr. JOSEPH EVANS for their translations.

Köln, August 1969 K. J. ZÜLCH

Table of Contents

A. Biography of Otfrid Foerster 1

B. Epilogue 17

C Excerpts from Foerster's most Important Papers 25

I. Contributions to Neurological Analysis, Diagnostic and Clinical . . . 27

1. Early Papers on Coordination, Associated Movements, Contractures, Spasticity 27

2. Studies on the Anatomy and Physiology of Muscles and Peripheral Nerves 32

3. Foerster's Contributions to Orthopedics and "Physical Terapy" (Neurological Rehabilitation) 35

4. Clinical Syndromes 38
a) The Pallidal Syndrome 38
b) The Athetotic Syndrome 38
c) The Choreoathetotic Syndrome 38
d) The Cramp Syndrome 38
e) The Atonic-Astatic Syndrome 38
f) Transection Syndromes of the Spinal Cord 38

5. Neuroradiological Experiences 43

II. Contributions to Neurosurgery 45

1. Posterior Root Section for the Diminution of Spasticity ("Foerster's Operation") 45

2. Posterior Root Section in Gastric Crises 46

3. Section of the Pain Tract in the Spinal Cord, the So-called Antero-Lateral Tract Section 48

4. Spinal Cord Tumors 50

5. Surgery of Peripheral Nerves 52
6. Brain Surgery: Elimination of the Meningo-cerebral Scar 53
7. Operative Treatment of Brain Tumors 54
8. Operations on the Vegetative System 55

III. Contributions to the Applied Physiology of the Nervous System 57
1. The Dermatomes 57
2. The Sensation of Pain and its Conduction Pathways 59
3. The Extent of the Principle of Localization 62
4. Contributions to Neurophysiology 68
a) The First Electrocorticogram of a Brain Tumor 68
b) Hyperventilation as a Method to Provoke Seizures 69
c) Reflex Physiology 70
d) FOERSTER's Opinion on the Activating and Inhibiting Systems of the Brain Stem 71
e) Contributions to the Knowledge of the Vegetative System 72
5. FOERSTER's Interest in the Physiology of Special Senses 74

IV. Congress Speeches and Biographical Appreciations 77

Epilogue 81

Bibliography 83

Figures and Photographs 97

A. Biography of Otfrid Foerster

Youth:

OTFRID FOERSTER was born on the 9th of November, 1873, in Breslau. His father had advanced from grammar school teacher to the position of full professor of "Classical Philology, Archeology, and Eloquence". He was a man of superior erudition and spoke Latin as a living language, as did so many members of his generation. It may well be that it was from his father that OTFRID FOERSTER inherited his profound gift of eloquence. He passed his "Abitur-Examination" in 1892, having been consistently one of the best students in his class. History, classical languages, and mathematics were his favorite subjects, and in mathematics he far outranked his classmates. He was said to have been curiously awkward in composition and perhaps even bad in expression in his native language. This is astonishing in the light of his later achievements in writing and rhetoric. He had artistic tendencies as well. He taught himself to play the flute. He loved theater and particularly enjoyed comedies. In his later life he enjoyed occasional burlesque.

The Student:

Because of his special gifts his teachers encouraged OTFRID to study philology; however, he chose medicine and the natural sciences. Even at an early age he had taken pleasure in discovering nature. As a child he collected butterflies, plants and rocks. What persuaded him, in his second semester, to choose medicine as his life work is unknown to us. He studied in Freiburg, Kiel, and Breslau from 1892 to 1896 and he passed his State board examination in Breslau in 1897. As a student he frequently attended parties and was said to have been an excellent dancer. Despite his interest in social life, work early began to dominate his thinking. He passed his examinations brilliantly. In one of them the physiologist, HEIDENHAIN, regretted that even the best possible classification "very good" did not do justice to him.

The Assistant:

As a student FOERSTER worked in the Mental Hospital Leubus where the great psychiatrist KRAEPELIN had been Acting Assistant Director. As an undergraduate he also came in contact with WERNICKE in Breslau. These associations may well explain his interest in Neurology.

Upon WERNICKE's suggestion FOERSTER went abroad for two years after finishing his thesis. He spent the winters in Paris with DEJERINE. He also met PIERRE MARIE and BABINSKI. During the summers he observed the physical therapy of neurological patients under FRAENKEL-HEIDEN of Switzerland and it was with FRAENKEL-HEIDEN that he published his first papers in neurology. It is noteworthy that FOERSTER was involved at the beginning of his medical career in the solution of the problems of "practical therapy" of neurological patients, but his unusual intellectual gifts of an analytical character drew him later into close contact with the basic sciences. Therapy at this time was in its infant stages. He tried to work out some scientific basis of practical therapy and never lost interest in this theme. Today "physical therapy" maintains a central position in rehabilitation.

During this period of his life he was never a stranger to his social environment. He enjoyed his stay in Paris. He was quite popular at parties, played tennis well, and was known for his elegant appearance. While a student in Paris he greatly enjoyed society and the theater.

On returning to Breslau to study with WERNICKE in preparation for his professional qualification, work became his dominant though not all consuming interest. He met his future wife, MARTHA, then sixteen years of age. He often played tennis with her and with her sisters. In winter everbody went skating together and OTFRID took the girls to the various balls. He soon became engaged to MARTHA, a known beauty; however, the engagement was not announced until after his qualification. After a year and a half of marriage a daughter was born who died in infancy, quite suddenly and without antecedent illness.

The unexpected death of the child seems to have heralded a decisive turn in FOERSTER's life, affecting him deeply. He plunged into scientific work, perhaps just to forget. Never again did he share time freely in his family life, as he had formerly done, even after more daughters had been born. His friends opined that he had become introverted. His old, pronounced extrovert inclinations toward people and his surroundings may well have been buried with his first child. The old traits would occasionally ring through in the aging FOERSTER, especially while a host in his own home or in social life at meetings. Another factor in this change may have been that his proud conscience drove him relentlessly because he had gained international recognition of his work earlier than was granted to his peers. He lived solely for the hospital, the research laboratories, and for international scientific discussions, though time and again one notices with astonishment how diversified his interests had remained. In discussions of classical literature, on questions of geology, on the history of Silesia and many other subjects his basic grounding was clearly evident.

Physician and Scientist:

In order to assess properly FOERSTER's importance in science one must review the status of the Neurology of his time. The field had just begun to develop out of internal medicine

(Charcot, Erb, Gowers) and out of psychiatry (Wernicke, Meynert, Anton). Efforts were being made in internal medicine to separate neurological diseases, to describe them, and thus to shape a systematic nosology. Psychiatrists—if they had a leaning toward neurology— attempted to promote an analysis of function in the various neurological diseases and to correlate function and the morphological substrate (Broca, Wernicke, Dejerine).

Neurophysiology (Fritsch, Hitzig, Ferrier) was more concerned with the disciplines of physiology and surgery than that of neurology. Microbiology and serology, which could have clarified the infectious diseases of the nervous system etiologically and pathogenetically, were in their infancy. The Spirochaeta pallida had not yet been discovered. The old argument among Erb, Fournier, and Leyden concerning the specific origin of tabes had not been settled.

Foerster joined the ranks of the young functionally and morphologically orientated investigators in neurology who were interested in the new concept of "localization". He reflects clearly the influence of Wernicke with whom he had published an "Atlas of the Brain" in 1903. Descriptive neurology—known to him from his apprentice years in Paris —had little interest for him throughout his life.

I have already alluded to the theme to which Foerster dedicated himself in his early years and which continued to hold his interest throughout his life—physical therapy. Tabes dorsalis and its consequences, particularly tabetic disturbances of gait, made up at that time a large fraction of neurological disease. Drug therapy was as yet largely powerless, but the questions of treatment were manifestly urgent. "Physical Therapy" was the only avenue of help open to these unfortunates.

The question was frequently raised during his later life whether Foerster actually was "the great physician" or whether he should not be regarded as a great scientist and researcher. The question seems to me to be superfluous or even poorly put. Certainly Foerster was not the warm hearted and kind physician one might have wished for as the ideal, but practical therapeutic effort always remained a major portion of his work as a physician. He was a doctor always much sought after by his patients. One should not forget that his first study was dedicated to "Practical Therapy" in neurology and this was at a time when the Schools of London and Paris were largely satisfied with "diagnosis". It was only later that the true significance of the analytical part of Foerster's work came to be regarded as essential. It is easy to overlook the fact that for him pathophysiological analysis served as the approach to the solution of therapeutic problems. He knew from bedside experience that such analysis might lead to therapeutic suggestions, oftentimes of an operative nature. This is best exemplified in his studies on the problem of pain. He was able to give a host of suggestions leading to its operative elimination. Let us, however, return to our chronological description.

In FOERSTER's scientific life one subject followed logically upon another. Prerequisite to practical physical therapy was an analysis of disturbances in the sequence of motion, that is "coordination" (his professional thesis in 1902). Individual aspects of the motor and sensory systems were worked out by analysis. He wrote his first monograph on these results. His thinking demonstrated precise logic. The observation that the hemiplegia of the tabetic was not spastic called his attention to the significance of the spinal reflex arc in cerebral spasticity and suggested the possibility of treatment through interruption of the sensory arc. Thus he conceived posterior root section and recommended it to eliminate LITTLE's spasticity, spastic cerebral palsy. This procedure became known as "FOERSTER's operation" and very early brought additional fame to him in international scientific circles.

FOERSTER *as Surgeon:*

Having familiarized himself with posterior root section for the elimination of spasticity, he asked himself whether this procedure could not be used to interrupt afferent impulses in the tabetic suffering from lancinating pain in gastric crises. FOERSTER's posterior root section seemed to bring good results. The operation in the tabetic, the first of many suggested by FOERSTER for the elimination of pain, provided the occasion to study the areas of sensory supply to the skin and this led to the determination of dermatome borders. This was pioneer work in the field of neurology. It followed the early observations of HEAD and SHERRINGTON and was the first clinical experimental study of posterior root section in the human. The SCHORSTEIN lecture in London in 1932 was a conclusive report. The results were subsequently recorded in the Handbook of Neurology.

When, following root section, one of his patients developed intolerable pain as a complication, he ventured in 1912, together with TIETZE, to transect the anterolateral column carrying the pain pathways within the spinal cord. This he did independent of SPILLER and MARTIN, even though a year after their publication. This procedure remains the standard method for the operative elimination of pain refractory to drugs. Later, with GAGEL, he published a study of the pathophysiology of the anterolateral funiculus, a paper which today represents one of the fundamental investigations dealing with this problem. These operations performed with TIETZE familiarized him with the scalpel and he did not hesitate in World War I to take up the surgery of peripheral nerves. In his opinion, peripheral nerve surgery did not receive sufficient attention from the general surgeons and their technique seemed too rough. Unlike anyone before him, he acquired an experience in hundreds of nerve sutures and neurolyses. Toward the end of the War he took over the operative treatment of spinal cord injuries and carried out independently the operative treatment of gunshot wounds of the head. The scalpel, successfully mastered by him during the war, was not handed back to the general surgeon on the return of peace. In 1917 he reported the successful removal of an intramedullary tumor of the spinal cord. His report in 1920 dealing with 12 cases of spinal cord tumor in nine of which there followed considerable restitution of function created a sensation. The paper contained perceptive comments on diagnosis and pathophysiology. He pointed out a reliable method of examining epicritic sensibility and its zones of disturbance, drawing attention to the value of figure writing on the skin.

Thus the neurosurgery of war time was followed by a transition to the problems of civilian practice. Among these was the operative care of patients with scars resulting from brain injury. Convulsive seizures were noted to increase in number in individual patients and FOERSTER tried to reduce the frequency of attacks by excision of the scar, which he postulated to serve as a stimulus for the seizure. His classical investigation of epilepsy through a functional analysis of the cerebral cortex was based on hundreds of cortical stimulations carried out during operation.

Every surgical procedure was utilized by FOERSTER to gain information that could not be procured otherwise and the information gained inevitably led to improved diagnosis. From this experience he gained a mastery of the anatomy and physiology of the central nervous system, which gave him an advantage over all other early German neurosurgeons.

It is not surprising in view of his familiarity with operations for meningo-cerebral cicatrix that he was not hesitant to undertake the removal of brain tumors. During the third and fourth decades of this century he carried out many operations for brain tumors with a degree of success that far exceeded that of the general surgeons. For example, he succeeded in removing a tumor of the quadrigeminal plate, the world's second case, a success that is the more surprising because his technique left much to be desired. He had never undergone any general surgical training and in his operative exposure of the nervous system lacked methodical exactness. The absence of elegance in his surgery of the "external coverings" was more than compensated for by his gentle and delicate way of working as soon as the cortex was exposed. Curiously, he did not keep up with the rapid development of neurosurgical technique in the decade and failed to accept any of the numerous methods which could have facilitated his operating.

His sojourn at CUSHING's clinic in 1930 where he was appointed "Surgeon-in-Chief, pro tempore" at the Peter Bent Brigham Hospital exposed him to the technique developed by CUSHING; however, after his return he did not change his operating habits. He kept on working without silver clips, without electric drill, without suction or electrocautery, and sometimes with an awkward and disadvantageous positioning of the patient on an old operating table. The patient was under local anesthesia and fully conscious. One wondered constantly how, in spite of all these deficiencies, FOERSTER could stop every bleeder with ligatures, pieces of muscle and with sponges. His narrow and skilled hands were enclosed in rubber gloves covered in turn with cotton gloves, which must have impeded delicacy of touch.
FOERSTER watched the development of the upcoming German neurosurgery with sympathy. He contributed a foreword to the first neurosurgical journal, the "Zentralblatt für Neurochirurgie" edited by TÖNNIS. He offered for the first issue a paper entitled Ependymoma.
The highlight of his life as a neurosurgeon was the visit in 1937 of the British Association of Neurological Surgeons in Breslau. The Society, coming from the TÖNNIS Institute in Berlin, listened to his extended presentation on the diagnosis of brain tumors offered in fluent English without recourse to notes. He subsequently was accorded the highest honor of the Society, a nomination as Member Emeritus.
FOERSTER was a great examplar of the value of being both neurologist and neurological surgeon. MONRAD-KROHN felt that only a Titan like FOERSTER could do justice to both disciplines.

FOERSTER *and the Basic Sciences:*

During his stay in Paris FOERSTER had become particularly impressed by the analytical and morphological approaches to the peripheral representation of the central nervous system. He had been exposed to these methods by both DUCHENNE and DEJERINE. These approaches, however, did not penetrate French Neurology. Both CHARCOT and PIERRE MARIE tended to be occupied by the clinico-semiological approach with its emphasis on case description. This explains the failure of development of a rapport between FOERSTER and French neurology. In contrast he was very much impressed by English neurology and neurophysiology, particularly by the work of HUGHLINGS JACKSON and the research of SHERRINGTON, whose *Integrative Action of the Nervous System* was referred to by FOERSTER as one of his Bibles.
FOERSTER as a neurophysiologist both in theory and in methods was transcendent,

particularly after he had succeeded in winning over the gifted ALTENBURGER as a co-worker. FOERSTER and ALTENBURGER were the first to describe the localization of a brain tumor by electrocorticography and they soon after reported on thirty patients with various problems in whom the new method of analysis had been applied. FOERSTER's ideas concerning the function of the anterior and posterior roots and of the various tracts of the spinal cord and their ultimate distribution to different parts of the brain were subjected to neurophysiological scrutiny.

By the 30's FOERSTER's name had become widely known in the Anglo-Saxon world and he began to attratc young American neurologists and neurosurgeons to his clinic, just as in earlier years young men had made an effort to spend some time in the SHERRINGTON laboratories. FOERSTER's close relationship with Anglo-Saxons was reflected in his consideration of SHERRINGTON as one of the greatest living scientists. SHERRINGTON reciprocated in his esteem of FOERSTER as one of the great neurologists of his time. SHERRINGTON wrote before his death that he regarded FOERSTER as worthy of the Nobel Prize for presenting clinical neurology in its entirety on the basis of a comprehensive knowledge of the physiology of the nervous system. In 1932 FOERSTER gave the SCHORSTEIN lecture on the Delineation of Dermatomes. In 1935 on the One Hundredth Anniversary of JACKSON's birthday, he was presented with a golden Jackson Memorial Medal, the highest honor to be accorded to any neurological scientist of the world at that time. On this occasion he presented his formal address on "The Human Cerebral Cortex in the Light of JACKSON's Teachings".

A memorable phase in FOERSTER's life was the call to LENIN's bedside upon the recommendation of the German ambassador, BROCKDORFF-RANTZAU, to the Foreign Minister, RATHENAU. He spent the time from June, 1922 to January, 1924 with occasional interruptions at his bedside. On FOERSTER's recommendation, BUMKE was asked in consultation as also were for shorter intervals MINKOWSKI, STRÜMPELL and NONNE. Though this episode removed FOERSTER from clinical work he nonetheless found a ready substitute in the pursuit of scientific work.

His mission had great political significance at that time. He, himself, considered it as a personal gain to have known this extraordinary man so thoroughly during his illness.

We have outlined above some of the major directions of FORSTER's scientific life and only scant reference has been made to his interest in clinical problems. Parkinsonism, observed more frequently after the First World War, led to his great study on the pathophysiology of the basal ganglia, a subject of interest then and still open to discussion today. He was one of the first to report on encephalography, then popularized by BINGEL. With GAGEL he wrote a series of papers, describing very carefully the several kinds of tumors of the brain and spinal cord, his primary interest being in gangliocytoma. In his last years he interested himself particularly in the pathophysiology and the surgery of the vegetative nervous system, analyzing the pathways for the control of circulation, respiration, etc., subjects of frequent reports at the Wiesbaden Congresses of the German Society of Internal Medicine. A special style of work characterizes all these scientific investigations to which style allusion will be made later.

FOERSTER's *scientific Facilities:*

For decades FOERSTER financed all his scientific work himself, possible out of his extensive practice. Fate never favored him and he had to fight hard for his achievements. For many years his scientific efforts were carried out in the basement of the Wenzel-Hanke Hospital or in screened off parts of the large day rooms of the wards. It was not until 1932, when it was almost too late, that the Rockefeller Foundation built for him a modern institute, which was supported jointly by the City, the Province of Silesia and the University. In spite of frustrations FOERSTER turned down all calls to go elsewhere, to ERB's Chair at the University of Heidelberg, or to work with OSCAR VOGT at the Kaiser Wilhelm Institute for Brain Research in Berlin-Buch. He maintained with VOGT a close scientific relationship while both were investigating the functions of the cerebral cortex—VOGT in the monkey and FOERSTER in his operations for the removal of cerebral cicatrix in the human. Both men agreed to carry out the same stimulation experiments on various portions of the cortex and they stipulated that the results were to be put down in writing and to be mailed on the very day of the procedure. Neither man wished to be influenced by the observations of the other.

However, despite their common scientific interest, a collaboration under one roof would hardly have been possible. These two rough-hewn personalites would never have gotten along amicably, desirable though this would have been for the sake of Science.

The Neurological Research Institute, which after FOERSTER's death came to be known as the Otfrid Foerster Institute, fell to the direction of VICTOR VON WEIZSÄCKER, with some resultant alteration in its scientific guidelines. The Institute was destroyed toward the end of the Second World War and Germany has not had again a Neurological Research Institute such as FOERSTER had built with departments for morphology, neurophysiology, and neurochemistry.

FOERSTER's *Position in German Neurology:*

FOERSTER's leadership in German clinical neurology was secure by 1924. FOERSTER and

NONNE were the intellectual leaders. FOERSTER, after NONNE's retirement, continued for eight years until 1932 as the President of the Society of German Neurologists. The splendor of German Neurology of that periods has never again been reached. WEIZSÄCKER has correctly stated in his Epilogue that " at the time that FOERSTER and NONNE directed the Society there originated a structure of congressional proceedings comparable to the structure of an organism". It was difficult for other societies to come up with anything to match. The Congress of German Neurologists owed its high standards in great degree to the personality and achievements of OTFRID FOERSTER. NONNE and FOERSTER together, the one's use of the method of case presentation teamed with the other's functional-analytical and therapeutic emphases, combined in a fashion that was very fortunate for the specialty. The reports of the meetings indicate clearly how FOERSTER imprinted his style upon the proceedings during his years as President, particularly through his addresses, which show him as an eloquent scientist of classical education at the peak of his performance and self confidence.

This development ended abruptly in 1934 when the Society was dissolved for political reasons and was merged with the German Society for Psychiatry. At that time German Neurology lost much through emigration of many colleagues who had lent wide support to the speciality in their positions as directors of neurological institutes or of city hospitals. After this loss the Congress of Internal Medicine in Wiesbaden became the forum for FOERSTER's scientific endeavor.

On one more occasion German Neurology bespoke its high standards when FOERSTER and BUMKE issued their Handbook of Neurology. The spirit of the book is FOERSTER's as is the style. It offered him the unique occasion to present his life's work in a concise form. The experiences of his lifetime are recorded with unbelievable accuracy, documented by a host of case reports supported by his flawless memory and careful records, combined with an unmatched review of the literature. He edited the five large chapters of the Handbook on the "Motor Areas of the Brain Cortex", the "Sensory Areas of the Brain Cortex", "Function and Innervation of Muscles", "On Physical Therapy", and the chapter on "Structure and Function of the Spinal Cord", a study which was particularly unusual and comprehensive for its time. In this way he supplemented the chapters already published in the first Handbook (LEWANDOWSKI), dealing with the "Periphery", which he had written during his sojourn in Russia at LENIN's bedside and which were based upon his experience with injuries of peripheral nerves. The supplementary volumes of LEWANDOWSKI's First Handbook of Neurology remain masterpieces, dealing with the anatomy and physiology of muscle and of peripheral nerves. The fourth volume on

"Injuries to the Spinal Cord", which was also published at that time, is less well conceived.

One may ask today whether FOERSTER's life and work reached full fruition. Rudiments of a major work on brain tumors remained unpublished, but in retrospect one may say with critical judgment that with his particular approach he would not have been able to do further pioneer work of the qualitiy that characterized all his scientific publications. He had been able in the decade between 1927 and 1937 to preserve for posterity the entire breadth and horizon of his knowledge, experience, and achievements.

FOERSTER *and International Science:*

It is surprising, perhaps owing to peculiarities in FOERSTER's personality, that he failed to create a "school" in Germany. Clinical collaboration with his gifted colleague, SCHWAB, ended early with the death of the latter. SCHWAB's successor, LUDWIG GUTTMANN, had to leave Germany in 1933 because of political developments. ARIST STENDER, pupil of NONNE and BAILEY, raised the standards of neurosurgical technique in Breslau to modern levels. He became FOERSTER's successor as neurosurgeon. However, FOERSTER, unlike NONNE, did not leave behind a large number of disciples who as collaborators or even trainees of collaborators occupy most of the chairs of German academic neurology today. In spite of this fact, his sphere of scientific influence far exceeds that of NONNE. Current Anglo-Saxon neurophysiology and neurology, particularly the American variety, are unthinkable without FOERSTER. The close friendship between FOERSTER and CUSHING was reflected in the scientific work of each. SHERRINGTON and FOERSTER as intellectual teachers merged in FULTON who, in turn, profoundly influenced American neurophysiology. PENFIELD, who worked for a time with FOERSTER in Breslau, carried on the analysis of the localization of cerebral function and of epilepsy in systematic fashion. PERCIVAL BAILEY too reflected FOERSTER's influence and in turn brought the new classification of gliomas to Breslau. He dedicated to FOERSTER the first edition of his book on brain tumors. Numerous neurosurgeons and neurologists holding academic chairs in North America carried with them decisive ideas derived from their stay with FOERSTER in Breslau.

FOERSTER's *Significance for the Science of this Time:*

What is the significance of OTFRID FOERSTER for neurology in the century in which he lived? Three points can be emphasized: 1) enrichment through singular neurological experience, 2) a new style of work, and 3) new conceptions concerning the organization of the central nervous system. Let it not be held against me if I try to recapitulate FOERSTER's most important scientific contributions and to catalogue them. In their sum total they show the greatness of his work. He actively investigated the structure and function of the pain system and determined the importance of individual components in the creation of the experience of pain. We owe to him numerous operations for the modification of intractable pain, particularly procedures directed toward the autonomic nervous system. He described accurately the topographical and segmental configurations of the peripheral nervous system. He carefully determined which muscles depend on what nerve supply and he showed that after injury to peripheral nerves, careful pre- and postoperative care may lead to restitution of function. He demonstrated that surgery may be required in a certain number of cases.

He introduced improvements in the techniques of the surgery of spinal cord tumors and showed the importance of excision of the meningocerebral cicatrix in the therapy of posttraumatic seizures. Other than FEDOR KRAUSE, who came up through the ranks of general surgery, he was the only successful German neurosurgeon of his time. A neurophysiologically oriented clinician, he developed our present ideas on basal ganglia and transverse spinal cord syndromes by means of basic analysis. The atonic-astatic syndrome is named after him. He found that a latent convulsive tendency can be activated by hyperventilation, a method that has become a standard way of provoking "seizures" during electroencephalographic studies. The term "psychomotor epilepsy" was coined by FOERSTER. The first direct corticography was carried out in his clinic as a part of his study of brain tumors. We owe to him our knowledge of the dermatone borders. He was the first to point out the importance of physical therapy in neurological illnesses, an insight which even today is only slowly permeating neurologicial thinking and which has become the basis of rehabilitation. Finally, his numerous clinical experimental papers should be mentioned in which he, GAGEL, and ALTENBURGER tested certain clinical or pathological questions against operative results—all procedures designed to improve therapy. We owe to his work an accurate analysis of spinal cord, brain stem, and midbrain function.

The second striking characteristic of his neurological work is its style. FOERSTER for the first time formed a type of "Neurological Institute" within whose confines problems were analyzed in greater detail through the collaboration of morphological and physiological teams, basic science workers and clinicians. A separate department of Neurochemistry was planned. His special modus operandi requires comment. The starting point of any scientific project was always an observation at the bedside. There was then formed a provisional working hypothesis to be tested clinically during operation. If possible the hypothesis was tested neurophysiologically as well. Whenever this was impossible, the problem was reproduced in an experimental animal and investigated in that way. Finally, there came the morphological control of the organ itself. The result would then be reflected back to the bedside for diagnostic and therapeutic purposes. A close relationship between clinical and basic scientific work characterized the Breslau style of that period. Similar work had been done by others before him. One thinks, of course, of HUGHLINGS JACKSON, the ingenious English neurologist. His work was firmly oriented philosophically, but was practically without the control of basic science. It represented a kind of anthropological neurophysiology. One thinks also of SHERRINGTON, but his neurophysiology was based solely upon animal experiments and remained to be tested clinically. It was through FOERSTER's influence that neurophysiology became oriented toward the clinic. His influence was most evident in the change of direction of American neurophysiology, particularly as seen in its most prominent representative, FULTON.

The third facet of his work which demonstrates FOERSTER's significance for the science of his century relates to his impact on our image of the organization of the nervous system. Strongly influenced as he was by the ideas of the great English neurologist, JACKSON, the distinguished intellectual pupil of the philosopher, HERBERT SPENCER, he developed a system of thinking based upon exact morphological and physiological foundations. His concept of the nervous system envisioned the collaboration of many parts and many levels. Each part had to fulfill its particular task and thus contributed to the total output. FOERSTER's use of the concept was exemplified in his analysis of the activity of the cerebral cortex. The pyramidal area, the extra-pyramidal, and the socalled adversive areas share with the deep basal ganglia one common effort, which breaks down if one part of the union suddenly drops out, notwithstanding the integrity of the remaining members. Failures of function were demonstrated by clinical examples. Though eventually one might witness reorganization based on the resumption of cooperation by the remaining parts, the restored function was by no means normal, the degree of recovery depending on the importance of the part eliminated from the common effort. Characteristic disturbances always remain related to the special function of the eliminated part. He endeavored to buttress this concept through new observations of increasing precision and he may have gone too far in attempting to prove his dogma. Individual parts were put together more and more precisely in a magnificent mosaic of function in which every part had its seemingly well known and secure place. It finally appeared as if the brain and spinal cord functioned simply as a technical machine. This was his depiction of the localization principle as he outlined it in his presentation to the German

Society of Internal Medicine in 1934. The presentation itself was a masterpiece of thinking and a unique example of functional analysis. Unfortunately, there ensued no intellectual argumentation with his greatest adversary in this line of reasoning, VICTOR VON WEIZSÄCKER.

Let us then pose the question: How much further have we progressed in our knowledge since FOERSTER's death? Certainly, the contributions of individual components to the functioning of the brain, their composite relationships, the effects of stimulation and inhibition are now much better known and are more precisely defined, thanks to new methods of electro-anatomical research. Certainly, neurophysiological techniques have been improved in the interim, permitting the simultaneous pickup of electrical currents, representative of function, at numerous points of the brain from different depths, permitting one to analyse the spread of the stimulus. With microelectrodes we can today even pick up the electrical activity of individual cells. However, in our total concept, we have not progressed very far beyond FOERSTER. Even the significance of the *substantia reticularis* as a well contained system of stimulation and inhibition was recognized clearly by FOERSTER. It seems to me that OTFRID FOERSTER was a researcher congenial to HUGHLINGS JACKSON, whom he so worshiped and for whose birthday anniversary he gave the memorial address on the Organization of the Nervous System.

Further Biographical Commemorations of FOERSTER:

In Memoriam OTFRID FOERSTER. Dtsch. med. Wschr. *79*, 55–56 (1954).
Memories of OTFRID FOERSTER. Zbl. Neurochir. *14*, 286–292 (1954).
OTFRID FOERSTER 1873–1941. Große Nervenärzte, Vol. I. Ed.: K. KOLLE
OTFRID FOERSTER and the Breslau Medical Faculty. In: Yearbook of the Schlesische Friedrich-Wilhelm Universität of Breslau. Würzburg: Holzner Press 1963, pp 316–388.

B. Epilogue

A series of obituaries have depicted the life and person of OTFRID FOERSTER very vividly and shown the significance to the world of his scientific work. Here we want only to cite the words of his successor, V. v. WEIZSÄCKER, who has expressed beautifully his appreciation of his predecessor's worth.

Victor von Weizsäcker:

OTFRID FOERSTER 1873–1941 *

. . . The early papers from Foerster's hand show, as is so often true of authors, all the traits that were to remain lifelong characteristics. Foerster moves from the nosological description of neurological symptoms to observation interpreted in terms of function, observation invariably based, however, on morphology. The determination of function is for him a localization. This seems to be the cardinal rule of his life's work.

In Breslau one still sensed the spirit of the physiologist, Heidenhain. Foerster, like Carl Ludwig and even Sherrington, belonged to the old school of physiologists who were almost equally anatomists, correlating physiology with structure. It is understandable that Otfrid Foerster, oriented intellectually toward physiological anatomy, now becomes a pupil of Carl Wernicke. He admired and worshiped his teacher. Wernicke remained a standard for him.

. . . These youthful papers, dealing with symptom analysis, display clearly the nature of Foerster's thinking. A most accurate physical examination is related to a certain image of structure and function of the nervous system. Wherever the influences of schooling and not those of a natural mode of thinking are determinants there is first evident the influence of the French Neurological Clinic, particularly represented by Duchenne (de Boulogne) and Dejerine and then the teachings of Hughlings Jackson, that "prophetic genius", the author of the "Bible of Neurology" as Foerster puts it. He spent two years studying in Paris, but the fascination of Anglo-Saxon neurology, permeated by Jackson's spirit, provided him with a new method of scientific attack . . .

. . . And so in 1909 Foerster in collaboration with Tietze and Küttner began to treat serious cases successfully by posterior root section. At that time the spinal cord proper was still a "Noli me tangere". When recurrence of pain appeared, they went on to high cordotomy.

Through these steps Foerster became a pioneer in a new field, neurosurgery—steps that had been taken before him by Victor Horsley and Harvey Cushing. A fateful element entered Foerster's life at this stage, for he was a man whose gifts qualified him to be a pure and wholly dedicated research scientist.

However, he was first a physician and then, in addition, a universally oriented natural scientist. The choice of neurology as a professional activity, a discipline which had never

* Dtsch. Z. Nervenheilk. 153 (1941).

occupied a central position in the German academic organization, contributed, no doubt, to the development of energies of explosive force in this extraordinarily singular and ambitious man by the sheer reality of his situation. He had no choice but to reach over into surgery and orthopedics, internal medicine, pathology, and into the field of medicine in its entirety. It is an important and interesting fact for the biographer that FOERSTER, the neurologist, functions as natural scientist, pathologist and neurosurgeon. The image of the modern neurologist was a unique creation in the person of FOERSTER. In 1913 in the Neurologisches Zentralblatt, FOERSTER initially indicated that he let no spinal cord operation pass without utilizing the opportunity, through electrical stimulation of the roots in the human, to procure topographical data on spinal innervation. FOERSTER's new style is here clearly demonstrated for the first time: reporting on numerous individual observations in the most concise form and demonstrating the experimental anatomy of segmental innervation of muscles in the human. There then followed papers employing photography extensively to serve as protocols of his observations. FOERSTER became the master photographer in neurology. Numerous early pictures have been preserved. He holds the patient and guides him as an artist holds his cello. Never thereafter can we read in his finely outlined but still youthful face his zeal and his unbroken joy in life. Then came the World War ...

... And then the career of the neurosurgeon was determined. In 1920 when the "Gesellschaft Deutscher Nervenärzte" again convened, one of the sensations of the meeting was his achievement of the return of function in nine cases out of twelve spinal cord tumors, one of them an intramedullary glioma. We must recognize that FOERSTER oftentimes ventures into the unknown and that he tests to its limit the tolerance of the organism against the efficacy of surgical treatment. He is a venturing pioneer whose inroads penetrate into untrodden, unsurveyed land. The history of surgery knows of the tie between its achievements and such daring courage, but one must recognize that the advantage of FOERSTER's neurosurgery over that of other surgeons who have ventured to approach "this most subtle of all organs" was that they lacked his splendid mastery of the anatomy and physiology of the nervous system. They sometimes even lacked respect for the infinite delicacy in the assigned tasks of various portions of the nervous system. Every liver and every lung cell has an identical function. In the nervous system probably no one element has the same assignment as another. It is understandable that FOERSTER's achievements, and even the necessity of a separate specialty of neurosurgery, went unrecognized by many surgeons ...

... From now on FOERSTER is the center of a growing sphere of action, the radius of which soon extends beyond the borders of Germany. I know of no clinician who offered with such unimaginable diligence so many meaningful papers while carrying on such demanding clinical, practical and operative work, remaining meanwhile a searcher of such perseverance and passion. He was granted personal gifts and gratification beyond those normally experienced.

... FOERSTER never missed an opportunity during an operation on the nervous system to establish by electrical stimulation the function of an exposed part. He thus developed an "anatomy of the living", progressing from the individual muscle to the roots of the spinal cord to the "tract system" of the cord and on to the cerebral cortex.

What DUCHENNE, BROWN-SEQUARD, HITZIG, FERRIER, JACKSON and many others had initiated is thus brought to a conclusion. The concepts that originated in animal experiments on localization are now completed in the human. The myeloarchitectural research of MEYNERT and FLECHSIG, the cytoarchitectonics of BRODMANN, VON ECONOMO, and of O. and C. VOGT have thus become related to experimental results in man. Thereafter the experimental analysis achieved through physiological stimulation is supplemented by a selection of cases of pathological loss of function resulting from the destructive lesions of trauma, surgery, inflammation, neoplasia, etc. Thus the experimental investigator and the operator stand next to the experimenting illness ...

... It is a basic difficulty of any such experimental design that artificial stimulation produces conditions differing from those occurring under natural circumstances. If a part of an organism is destroyed, we see only how the rest of it works when the destroyed part is missing. Therefore, neither the observation of sequelae of stimulation nor the observation of sequelae of destruction permits a conclusion of certainty concerning normal function, particularly because the nervous system is capable of appropriating new functions. Thus it comes about that many functions, starting with that of speech, became "localized", although in reality one could only say that they had disappeared in certain lesions ...

... This principle extends even to the localization of consciousness, whether in the cortex or in the brain stem. FOERSTER escaped these dangers, especially characteristic of the WERNICKE school, by subscribing to the localization doctrine only insofar as he was able to construct it as an experimental science with particular reference to man.

Thus the scientist has put down the work of a lifetime in roughly 2,000 pages with about 1,700 illustrations, most of them his own photographs, photographs representing only a part of his collection, though the most important ones. One might call the result a living anatomy as well as an anatomy of the living; but one must add immediately that

Society of Internal Medicine in 1924. The presentation itself was a masterpiece of thinking and a unique example of functional analysis. Unfortunately, there ensued no intellectual argumentation with his greatest adversary in this line of reasoning, VIKTOR VON WEIZSÄCKER.

Let us then pose the question: How much further have we progressed in our knowledge [illegible] functioning of the brain, their composite relationships, the effects of stimulation and inhibition are now much better known and are more precisely defined, thanks to new methods of electro-anatomical research. Certainly neurophysiological techniques have been improved; in the human, recording simultaneously of electrical currents, representative of function at numerous points of the brain from different depths, permitting one to analyse the spread of the stimulus. With microelectrodes we can today even pick up the electrical activity of individual cells. However, in our total concept, we have not progressed very far [illegible] with the significance of [illegible] stimulus responses as well as [illegible] excitation and inhibition [illegible] by FOERSTER. It seems to me that FOERSTER's work [illegible] remained, [illegible] whom he so [illegible] in his memorial address on the Organisation of the Nervous System.

Further Biographical Commemorations of FOERSTER

[illegible] FOERSTER. Dtsch. med. Wschr. 79, [illegible] (1954).
[illegible] Zbl. Neurochir. [illegible] (1954).
[illegible] Dtsch. Nervenheilk. [illegible] 1954.
[illegible] and the Breslau Medical Faculty. In: Yearbook of the Schlesische Friedrich-Wilhelm University of Breslau. Würzburg: Holzner Press 1963, pp. 376–388.

the concept of function is limited unequivocally and sharply in two directions, "phenomena of stimulation" and "phenomena of lesions". Jacksonianism in neurology was never pushed to its limit so effectively and no one other than KLEIST applied WERNECKE's clinically founded localization principle in a fashion comparable to FOERSTER. One may say, however, that a fundamental distance between KLEIST and FOERSTER has not yet been bridged and in this polarity of basic ideas there is reflected a difference of type and gifts . . .

. . . This does not mean that there was nothing in common between the two men in their efforts to solve the problem of nervous function through localization. In spite of GALL's ill-conceived theories, it is FLOURENS who was the founder of experimental research on localization, even though universalism is erroneously attributed to him. SCHIFF, JACKSON, BROCA, FERRIER, HITZIG, WERNICKE, VOGT and FOERSTER followed in his path. However, any attempt to reconstruct by scientific, historical methods the origins of the term "irritability, irritation, conduction, inhibition, reflex, coordination, center, periphery"—and how they enter FOERSTER's system of neurology—is hardly feasible. In any event, FOERSTER parts from WERNICKE at the point where he selects for his method the experiment adapted to neurophysiology and where he analyzes natural, pathological phenomena as experiments of nature. If one asks whether he was more clinician or more theoretician, there may on first glance be some confusion. It would be easier, of course, to say that he was both. But what was the goal of his intellectual passion at the bedside, in the operating room, in the research institute, and at his desk? There is no doubt that it was the function of the structure and that is a theoretical problem . . .

. . . Equally characteristic is his approach to theory and to general discussion. He always submits to a fact of experience but never to a contrary opinion. You may say of him what he said about JACKSON: He was, as he says, neither a frantic localizer, nor was he a universalizer. But in his passionate practice FOERSTER was a localizer just as was JACKSON. This fact rarely appears expressly stated in his papers, but it was always brought out in live debate: "Professor WEIZSÄCKER thinks we can say nothing about the function of any single portion of the nervous system. If I ascertain that upon stimulation of the peripheral nerve, the muscle contracts and if I ascertain further that after section of that nerve, these muscles no longer contract on stimulation of the central nervous system, I do not see why we are not justified in saying that it is the function of the peripheral nerves to innervate the muscles. This terminology is centuries old and if we should have to give it up because it was declared untenable, I request that something be put in its place which I can understand and which we may discuss" . . .

... For FOERSTER the nervous system was truly a system, not only a structure but also a function. This is the central point in his thinking and it has significance also for the future. He is a research scientist, not only in ideation but also in experimentation. More correctly, he creates the experiment ...

... If there ever was an age for the natural scientist to be universal, this age is long gone. It antedates SOCRATES. Examples like LEONARDO and GOETHE confirm rather than contradict the statement. It is a requirement for all researchers to be restricted or they are no longer researchers. Without vigorous effort in one direction, no progress is realized in the natural scientific sense. When asked to remove and delete much that is cumbersome in order to make progress, FOERSTER makes use of this right of restriction. Furthermore, facts need to be associated. Light needs to be thrown upon them and interpretation is essential. Therefore, the scientist deletes what does not fit his basic concept so long as be does not eliminate or distort contradictory evidence. Above all, one finds what he is looking for. This methodology appears when FOERSTER tries to explain new observations, but he recognizes that deviations from the usual are to be expected ...

... FOERSTER's life was not spoiled with the modern helpful conveniences of demanding research. Everything in his environment was scarce and he always had to fight hard for it. He has proved that a beautiful institute, by itself, does not create the scientist, but that the institute comes because of he scientist. As is so often the case, favorable working conditions came almost too late, certainly later than he deserved. Work in the Neurological Research Institute commenced in 1934, the beginning of the last epic of his life's work. It was because of FOERSTER's orientation to the Anglo-Saxon type of science that the Rockefeller Foundation established the Institute and it was thereafter that the city of Breslau recognized the necessity to step in with a helping hand. We are grateful that his life does not contain any of the "Laboratory apparatophilia"—verbally or figuratively—which has in many places become the signature of scientific decadence. FOERSTER's sympathy for Anglo—Saxon science was for its realism, not for its so—called technical dimensions ...

... Until 1934, FOERSTER had achieved everything with his two eyes and his two hands. As soon as it proved possible to multiply these instruments following the opening of the Research Institute, he began to get more impersonal. From then on in many papers appearing under his name he served only as a collaborator however indefatigably and intently he may have promoted them. FOERSTER was—coming from WERNICKE—a clinical neurologist, but he was an almost self-taught anatomist, neurosurgeon, and physiologist ...

... The clock stands still, the hand falls. "Patriae scientiae inserviendo consumptus". The first three of these words, engraved in 1934 on the entrance of the OTFRID FOERSTER Institute in Breslau, are now supplemented by the fourth, fully revealing the meaning concealed to the present. One understands the supplementation by this last word. One knows enough. One knows everything about the man portrayed in this brief report on his life's work. OTFRID FOERSTER was buried on one and the same day in one and the same grave with his life's companion. The same disease carried off both. Such an end is worthy of his life. Seen geographically, FOERSTER is an Eastern German symbol of hard, tough, self-renouncing work. However, the work of his hands was the consuming work of his spirit.

Laurel belongs on his bier ...

Further Memorials

ALEJANDRO H. SCHROEDER: OTFRID FOERSTER. Anales del Institute de Neurologia, Uruguay, Bd. III (1941).

K. M. WALTHARD: OTFRID FOERSTER (1874—1941). Arch. Neurol. u. Psychiat. *62*, 401.

G. HOHMANN: Was verdanken wir OTFRID FOERSTER? Z. Orthop. *72*, 279—284 (1941).

O. GAGEL: OTFRID FOERSTER 1873—1941. Klin. Wschr. *20*, 799—800 (1941).

O. GAGEL: OTFRID FOERSTER 1873—1941. Arch Psychiat. Nervenkr. *114*, 1 (1941).

OTFRID FOERSTER, 1873—1941. Presse méd. v. 24. 9. 1941, S. 1051.

EGAS MONIZ, Lissabon; Nobelpreisträger 1949: Lisboa méd. *19*, 52—58 (1942).

MARGARET KENNARD—JOHN FARQUHAR FULTON—CARLOS G. DE GUTIERREZ-MAHONEY: OTFRID FOERSTER 1873—1941. J. Neurophysiol. *5*, 1—17 (1942).

CARLOS G. DE GUTIERREZ-MAHONEY: OTFRID FOERSTER (1873—1941). Arch. Neurol. Psychiat. *46*, 913—918 (1941).

H. PETTE: OTFRID FOERSTER (1873—1941). Münch. med. Wschr. *44*, 1183 (1941).

H. PETTE: OTFRID FOERSTER. Der Kämpfer um eine selbständige Neurologie. Gestalter unserer Zeit Bd. 4, Oldenburg: Gerhard Stalling-Verlag, S. 93—100.

M. PASTEUR VALLERY RADOT: OTFRID FOERSTER (1873—1941). Rev. Neurol. Bd. 73, 1—2 (1942).

K. H. BAUER: OTFRID FOERSTER. Chirurg *13*, 431—432 (1941).

A. STENDER OTFRID FOERSTER (1873—1941). Dtsch. med. Wschr. *44*, 1214—1215 (1941).

The Creation of the Otfrid Foerster Medal by the German Society of Neurological Surgery, 1953

The OTFRID FOERSTER Medal was established in 1953 by the German Society of Neurological Surgery and a lectureship was created in his honor in order to keep alive his memory. The first lecture was given by PERCIVAL BAILEY before the German Society of Neurological Surgery on the 22nd of August, 1953 in Munich. Two years later the Medal was given to LUDWIG VON BOGAERT, in 1960 to W. TÖNNIS, in 1964 to A. E. SPIEGEL, and in 1966 to PAUL C. BUCY.

C. Excerpts from Foerster's most Important Papers

I. Contributions to Neurological Analysis, Diagnostic and Clinical

1. Early Papers on Coordination, Associated Movements, Contractures, Spasticity

FOERSTER had learned the place of physical therapy in the treatment of tabetics from FRENKEL-HEIDEN. This association had stimulated in him a special interest in coordination, the analysis of which is a prerequisite for the understanding of purposeful exercise. In Breslau he supervised actively the physical therapy of the neurologically ill and he carried out the most exact analysis of the muscles from the viewpoint of motion and coordination. In this he tried to follow his master, DUCHENNE.

In motion he distinguished between the principle physiological movement and associated movements. He then began to study "pathological" associated movements, which study later played a decisive role in the analysis of syndromes of synergistic movements (WERNICKE-MANN). In addition, he analyzed the alterations of tone, particularly in the syndromes of spasticity and contracture, so burdensome to the patient. The operative relief of spasticity in these conditions by means of posterior root section soon made him internationally famous.

The attempt to eliminate disturbances of coordination and of spastic contracture through therapeutic exercises necessitated more intricate examination of "flaccid and spastic" paralysis, which he later analyzed in BETHE and BERGMANN's Handbook in 1927, where there are gathered all his earlier papers and observations.

These papers illustrate FOERSTER's attention to minute detail in individual cases. They also offer, in the form of a discussion with HOFFMANN on muscle reflexes, an example of the literary argumentation typical of FOERSTER—an argumentation that he carried on quite sharply at times during the middle epic of his life.

The Physiology and Pathology of Coordination

Jena: Gustav Fischer Press 1902.

Associated Movements in Healthy Individuals and in the Neurologically and Psychically Ill

Jena: Gustav Fischer Press 1903.

... It would be a sin of omission not to mention here the monumental work which J. B. DUCHENNE erected to himself and to science in his work on the physiology of movement ...

... Almost all of our purposeful movements, even those that seem the simplest, not to speak of the composite ones, permit the recognition of a composition of primary and associated movements. Thus, on opening of the closed fist, an extension of the fingers is the primary motion, flexion of the hand the appropriate associated motion; in closing

the eyes the approximation of the eyelids is the primary action, a slight upward rotation of the eyeball and a slight narrowing of the iris are associated movements. In lowering the eyes, not only the eyelids are lowered; conversely, in lifting up the eyes, there occurs a lifting of the eyelids and a furrowing of the forehead as well as rotation of the eyeballs. This widens the visual field upwards in a purposeful fashion.
In as much as all these purposeful associated movements occur under normal circumstances and pertain to normal movements, they may be termed "normal purposeful associated movements". These are opposed by a class of associated movements which may be termed "purposeless associated movements" ...
... We find them particularly when a movement or manipulation is to be carried out which is newly acquired. Mention may be made of the child who in learning to write screws up his face, turns his head, sticks out his tongue ... Secondly, we always find them when a motion is to be carried out with great force. For instance, when we push up a heavy load with one arm, the face contorts to a grimace, the muscles of the other arm contract, the fist closes; truly, at the height of the effort the entire body musculature goes into action to a variable degree ...

FOERSTER now describes pathological associated movements.
... In hemiplegic patients we quite commonly observe that they always extend the hand when they wish to flex the paretic fingers into the palm; moreover, if the weakness of the flexors be somewhat pronounced, flexion does not succeed at all in the hemiplegic without this concomitant extension. The patient cannot suppress this hand extension no matter how hard he tries. The healthy individual, of course, succeeds without any trouble ... In this extension of the hand we recognize immediately the purposeful associated movement brought about by the organism to heighten the power efficiency of the finger flexors ... Likewise, we see an extensive flexion of the hand developing as the closed fist is opened ... If a patient is supine and he is to flex the lower leg in this position, the following associated movements will be strongly dorsiflexed. That is accomplished by the superior force of the tibialis anticus producing the varus position (STRÜMPELL). It is noteworthy that this dorsiflexion of the foot may be pronounced when the intent is to flex the lower leg, even if the patient is totally unable to bend the foot voluntarily. Besides dorsiflexion of the foot, there occurs almost regularly another associated movement, namely, flexion of the thigh. This becomes evident as the femur and pelvis assume an angulated position and the hip is elevated ...
... When one strokes the sole of the foot rapidly with the nail, the big toe is known to

dorsiflex in afflictions of the pyramidal tract. The other toes either carry out the same motion or they all plantarflex. Infrequently, however, the entire lower extremity flexes simultaneous at the ankle, knee, and hip joints upon the same skin stimulus. This occurs even though it may be altogether impossible for the patient to carry out this motion voluntarily. The retraction of the entire leg is certainly a reflex movement in these cases. We cannot, however, properly regard this as an associated movement, but rather as a more extensive purposeful reflex movement of defense . . .

. . . We have seen what an extensive and widely branching system regulates the course of voluntary movements. We have also seen how numerous and complicated mechanisms interplay to bring them about. The associated movements are only one of the external signs of disturbed mechanisms. However, though their cause may be different in the special case, according to a difference of this or that individual mechanism, all stem from one and the same principle. By this principle the organism must command as many means of motion as possible if it intends to perform a movement. It innervates, *a priori,* too many rather than too few muscles; a choice of our means of motion is a searching process, sometimes with a certain lack of economy in evidence. Normally, this lust for waste of effort is restricted through inhibitory mechanisms; however, under pathological circumstances, it manifests itself in associated movements . . .

Constractures in Diseases of the Pyramidal Tract

Berlin: S. Karger Press 1906.

. . . Contracture is a subcortical fixation reflex, or expressed more correctly, it is an exaggeration of the normal fixation reflex, of the normal resistance which every muscle by reflex opposes to its elongation . . .

. . . This easily explains a number of the peculiarities of these spastic contractures which we mentioned above and clearly sets them apart from the contractures of shrinkage . . . Now we understand why all sorts of sensory stimuli such as faradic current, cold and injuries, worsen the contracture. In the case of these stimuli continuous centripetal irritations are transmitted to the subcortical centers. These are repetitively charged and reflect the input stimuli back to the muscle so that even the slightest attempt at elongation produces very rigid antagonistic tension in the muscle . . .

Flaccid and Spastic Paralysis

Handbook of Normal and Pathological Physiology, Vol. 10. Berlin: Springer Press 1927, pp. 893–972.

... In view of STERNBERG's very carefully executed experiments there can hardly be any doubt that the receptors for the tendon phenomenon are not situated in the tendons but rather in the muscle itself or in the periosteum. At least the tendons of the muscle may be replaced by a thread; tapping the thread has the same reflex effect as tapping the tendon. However, this does not disprove the existence of the tendon reflex, for it is possible that tapping the tendon stimulates an intradendineus receptor. The evidence which HOFFMANN cites against the existence of periosteal reflexes stands on extremely weak ground. It cannot be denied that in eliciting osteoreflexes there usually occurs a sudden short pull on the muscle. This is most certainly the case when we tap the distal end of the radius with the humerus dependent and the forearm half flexed and assuming an intermediate position between supination and pronation. When HOFFMANN says that there is no reflex attendant on tapping a bone without putting pull on the muscle in a longitudinal direction, i. e., when he says that with the forearm fixed, the reflex from the periosteum of the radius to the biceps brachii is absent, he is wrong. Even when movement of the tapped extremity is fully eliminated there is often an equally prompt jerk of the biceps. Furthermore, I cannot admit that the triceps should always jerk upon tapping of the distal end of the ulna, because of being subjected to extension. Very frequently the biceps jerks exclusively or predominantly although it is not elongated by tapping the distal end of the ulna. On the contrary, it is approximated. It is difficult to explain many osteal reflexes on the basis of the muscle jerking in a longitudinal direction. If upon tapping the zygoma or the nose or the forehead, the entire facial musculature twitches even on the contralateral side and the nuchal extensors contract; and if upon tapping the clavicle, the triceps or biceps, deltoid, pectoralis major, and perhaps even the finger flexors contract; if upon tapping the spina or basis scapulae and the deltoid, triceps, pectoralis major and other arm muscles contract; if upon tapping the epicondylus externus the triceps or biceps, the pectoralis major, the finger flexors contract; if upon tapping the dorsum metacarpi the finger flexors contract; if upon tapping the ribs or the superior iliac spina the abdominal musculature contracts; if upon tapping the epicondylus internus or externus femoris, or the anterior surface of the tibia, the quadriceps and the adductors contract; if upon tapping the dorsum of the foot the flexors contract; if upon tapping the major trochanter the gluteus maximus, biceps, semitendinosus and semimembranosus jerk; as it is frequently the case in the increased reflex activity of spastic paralysis; when, in fact, one cannot agree that the jerking muscles are subjected to tension in the lengthwise direction and that they should jerk for that reason, this is an ungrounded assumption, entirely. Some of the contracting muscles, on the contrary, even undergo an approximation of their insertion points. HOFFMANN, himself, recognizing the weakness of his arguments, speaks simply of percussion of the muscles rather than of extension. TROEMNER and KRAMER appear to accept HOFFMANN'S thesis ...

... HOFFMANN feels that contradictory observations have no weight. Certainly they do not if they do not fit into a doctrine well founded by serious arguments. And what single positive arguments do HOFFMANN and his supporters have in favor of the doctrine that bone reflexes originate only through extension or percussion of the muscle? Literally none. STERNBERG'S experiments cannot be gainsaid in a few words as HOFFMANN suggests. I have transected the posterior roots of L-2, L-3, L-4, L-5, S-1 on the paralyzed side in a case of severe spastic paralysis secondary to a traumatic lesion of the cortical leg area, in which case the tendon and periosteal reflexes of the paralyzed leg were very vigorously exaggerated and in which instance numerous muscles of the contralateral weak leg (quadriceps, adductors, flexor digitorum, etc.) could be made to contract by tapping the tendons and osseous prominences of the sound leg. After section of the roots, it was no longer possible to get quadriceps, adductors, gluteus maximus, biceps, semitendinosus and semimembranosus to contract by any stimulation of the tendons or bones of the paralyzed leg. The gastrocnemius reflex was still elicited weakly from the Achilles tendon as well as from the sole of the foot and the reflexes from the sole to the flexores digitorum persisted also. In this case the "fully deafferented" quadriceps on the paralyzed side could still be contracted significantly upon tapping the patellar tendon or the inner epicondyle on the contralateral side. The adductors of the paralyzed side twitched also. This observation speaks absolutely against the theory of traction-percussion and the thesis that all reflex effects emanating from tendons and bones are merely intrinsic muscle reflexes. The quadriceps and adductors on the paralyzed side had been fully "deafferented", their intrinsic reflex sensitivity had been removed; however, anterior horn cells had remained in undisturbed contact with the afferent pathways of the contralateral leg. Therefore, the muscles could still contract upon stimulation from nondeafferented areas ...

... These observations prove untenable that part of the thesis of HOFFMANN and his supporters that intrinsic muscle reflexes are responsible for the participation of the remote muscles in the reflex elicited from tapping tendons or bone. On the contrary it is evident that the contraction of these remote muscles originates from the activation of receptors in

the immediate target area, and through direct excitation of anterior horn cells of the remote muscles and of afferent impulses emanating from the site of the stimulus. It is evident further that the old thesis of the so-called reflex irradiation which HOFFMANN rejects as the explanation for tendon and osseous reflexes is still tenable.
HOFFMANN is further wrong in his thesis when he assumes that a muscle can participate in the reflex action only if its intramuscular sensory receptors are stimulated through extension of the muscles in a longitudinal direction or through percusion. Immediate coordinated reflex synergies may well be achieved by a tap on the tendon or on the bone. No stimulus better elicits the extensor kick of the leg, the coordinated stretch reflex of the leg, than a short tap against the sole of the foot. Percussion or stretching of the cooperating muscles (except for the gastrocnemius and the flexores digitorum) plays no part in the process ...

FOERSTER *Utilizes Movies in Teaching*

FOERSTER was the first German clinician to make extensive use of moving and still pictures.
Dtsch. Zschr. Nervenhk. *50*, 293 (1914).

... Then follows a movie demonstrating a rare disturbance of motion afflicting the right arm which might be interpreted as due to ataxia secondary to loss of position sense which really is due to apraxia. The case documents beautifully the absolute motor confusion of the right arm. The cinematographic demonstration closes with a case of HUNTINGTON'S chorea. The pictures were taken by the author using his own equipment and in cooperation with the photographer PROHASKA ...

2. Studies on the Anatomy and Physiology of Muscles and Peripheral Nerves

The detailed study of patients whith clinically demonstrable nerve injuries and the findings after root section endowed FOERSTER with a unique knowledge of the anatomy of nerves and muscles. He reported on the results in the first three supplementary volumes of the First Handbook of Neurology (LEWANDOWSKI). Later he enlarged the scope of his studies on the anatomy and function of the muscles and we owe to FOERSTER another contribution in the Second Handbook of Neurology. These four volumes on peripheral nerves and muscles have yet to find their match. They contain accurate description of the individual muscles of the human body, their innervation and variants thereof, as well as an exact analysis of their primary functions. Besides these analytical presentations, synthetic demonstrations of "movements" in individual joints are given, whereby the synergism of the muscles and their common action are described and the significance of the loss of any individual muscle is characterized. HOHMANN in his memorial note states quite justifiabl that even to this day an orthopedist can hardly get along without some knowledge of this chapter. Individual reports on the segmental and branch innervation of the muscles remain unique. Analysis of the structure of the cervicobrachial plexus as copied everywhere in textbooks is equally fundamental for the neurologist as well as the tabulations of the "nuclear columns" of the cord (1929, 939, 941 ff., 961 ff., 966 ff.) showing the levels of muscle innervation. The description of the abductor pollicis brevis is reproduced here as an interesting example from this chapter. According to FOERSTER, it is the only monosegmental muscle.

Physiology and Special Functional Pathology of Striated Muscles

Handbook of Neurology, Vol. 3: General Neurology III, p. 307 ff., 1937.

9. *Abductor pollicis brevis, flexor pollicis brevis, opponens pollicis, adductor pollicis.* (C8, Th1; *lower* primary medial fascicle, medial roots of the median nerve (ulnar nerve, musculo-cutaneous nerve—ulnar nerve).

. . . The *abductor pollicis brevis* is the most superficial of the muscles of the thenar eminence. It originates from the transverse carpal ligament, from the navicular bone, and from the tendon of the abductor pollicis longus, as a projection into its muscle bulk. Its tendon inserts into the outer aspect of the proximal phalanx of the thumb and merges into the dorsal aponeurosis of the thumb . . .

. . . Generally, the abductor pollicis brevis, opponens and flexor brevis are tributary to the median nerve, the adductor to the ulnar nerve; however, deviations infrequently occur in such fashion that the ulnar may innervate several and rarely all the muscles of the thenar eminence. Very infrequently the abductor brevis may derive some of its supply from the ulnar. Conversely, the median may contribute to the adductor pollicis. The overlap of the ulnar into the sphere of the median takes place by way of the anastomosis between the deep branch of the ulnar nerve and the median in the hand. The abductor brevis, flexor brevis and opponens are occasionally supplied by an anastomosis from the musculocutaneous to the median. The median gains part of the supply to the adductor pollicis by way of the anastomosis which springs from the volar interosseus branch, running obliquely down the forearm to the ulnar.

The abductor brevis, flexor brevis and the adductor are quite accessible to percutaneous electrical stimulation. The opponens, as a single muscle, can contract to its full extent only if the abductor brevis is paralyzed and atrophied. Actually, paralysis of the abductor brevis is, however, relatively frequent. As already mentioned, this muscle not infrequently is the only one to be supplied exclusively by the median so that in inter-

ruptions of the latter, it alone may be paralyzed. Moreover, it is often seen that following damage to the median, the abductor brevis restitutes as the last of the muscles supplied in the median distribution and remains the only one paralyzed. Furthermore, it is the only muscle of the upper extremity with monoradicular innervation. In isolated interruption of Th1 it is totally paralyzed, whereas all the other short muscles of the thumb and all small hand muscles remain functional because they receive concomitant innervation through C8 . . .
. . . The isolated action of the abductor pollicis brevis turns the first metacarpal in the dorsovolar axis of the carpometacarpal joint toward the radial side, while at the same time it undergoes flexion in the radioulnar axis and undergoes pronatory twist. Secondly, the proximal phalanx is tilted at the metacarpophalangeal joint to the radial side and is twisted into pronation, but is flexed very little. The terminal phalanx is extended . . .

Special Anatomy and Physiology of the Peripheral Nerves

Handbook of Neurology, Supplementary Volume, Second Section. Berlin: Springer Press 1929.

. . . Before the war the descriptive anatomy of peripheral nerves was presented as a seemingly closed chapter. It was almost dogmatically established what muscles are innervated by what peripheral nerve. Many demonstrations of abnormal motor innervations and nerve anastomoses offered by previous anatomists, particularly by LETIEVANT, on the basis of clinical experience, had not yet become common knowledge among neurologists. War injuries documented many discrepancies between the presumed total loss of function following section of a particular nerve and the observed retention of full or partial integrity of a number of muscles in the domain of that nerve. This has stimulated extensive neurological studies on the abnormal nerve supply to muscles. It is to be emphasized that for one muscle there may be an anastomosis between nerves so that motor fibers may transfer in small or large numbers, and for another a nerve may give off *direct branches* to such muscles as are normally supplied by another nerve. In this fashion a muscle may be innervated simultaneously by two nerves, or exclusively by a nerve normally not superordinated to it. This *double innervation* or *abnormal innervation* may explain preservation of muscle function following transection of a certain nerve. This also plays an important role in the case of so-called quick healing after nerve suture, of rapid return of function in certain muscles occurring because of the presence of anastomoses or anomalous branches. The muscles supplied by the anastomosis in such cases frequently do not retain their function, but rather regain it gradually, often after the performance of the nerve suture. Why there is a variation in the rate of return of function cannot be gone into at this point . . .

The Spinal Segmental Innervation of Muscles

Neurol. Centralblatt *19*, 1–14 (1913).

. . . I should like to list a personal observation. When in tabetics the 7th through the 10th thoracic roots (motor and sensory) were cut, there followed a total paralysis of the recti and oblique muscles down exactly to the umbilicus. Below the umbilicus, the function of the recti and obliques remained intact and no difference could be established between them. This finding fits those of SCHWARZ und speak against the assumption of GOLDSTEIN and IBRAHIM . . .

I am turning to the *lower extremity* now. My own investigations refer chiefly to the electrical stimulation of the individual anterior lumbosacral roots, which I was able to carry out in ten cases. In addition, I have used this technique for the accurate determination of the muscles paralyzed in diseases of the anterior horn cells in the lumbosacral region of the spinal cord . . .

3. FOERSTER'S Contributions to Orthopedics and "Physical Therapy" (Neurological Rehabilititation)

FOERSTER *and Orthopedics*

Because of his special interest in physical therapy and in the influence of surgery upon disturbances of movement, FOERSTER developed close relationships with orthopedists and at many meetings he spoke on orthopedic problems such as the elimination of spasms and contractures, torticollis, etc.

The Operative Treatment of Spastic Paralysis (Hemiplegia, Monoplegia, Paraplegia) Following Gunshot Wounds of the Head and Spinal Cord.

Dtsch. Zschr. Nervenhk. *58*, 151–215 (1918).

. . . Therefore, I should like to regard the sum total of all interventions on and for the hemiplegic leg as one operation. The steps in this operation are elongation of the *Achilles tendon, transplantation of a fascicle of the tibial nerve to the tibialis posticus and the flexor digitorum, splitting of the tendon of the tibialis anticus lengthwise and transplantation of the split portion to the lateral edge of the foot, and transplantation of the crural nerves to the individual heads of the quadriceps muscle . . .*

FOERSTER *Provided the Neurological Basis for Present Day Physical Therapy*

FOERSTER encouraged an interest in physical therapy among neurologists and orthopedists alike. He created among neurologists an interest in long-range therapy by careful analysis of "disturbed motion". He established the fundamentals for a purposeful gymnastic therapy adjusted to the individual situation, starting with the disturbances of coordination observed in tabes. He next emphasized pyramidal paralysis of random etiology. HOHMANN points out in his memorial note that FOERSTER laid the foundation for physical therapy in Orthopedics. Today it is quite evident that rehabilitation of motor function requires the closest collaboration of the neurologist because analysis of motion is essential; however, the prerequisite for improvement through therapeutic exercises requires not only knowledge of movement disorders, but also an accurate knowledge of what is possible, that is, recognition of the potentiality for restitution. Here, too, FOERSTER left fundamental precepts for the successful rehabilitation of the hemiplegic through his description of synergistic movements.

Physical Therapy in Tabes Dorsalis

Deutsche Ärzte-Zeitung 1901, Vol. 5.

. . . This therapeutic factor is *Exercise* for which reason the method may be correctly termed *Exercise Therapy. It* was employed for the first time systematically and on a scientific basis by FRENKEL, 11 years ago. Movements initially incoordinated become coordinated only through exercising; that is through their intentional repetition. Walking, jumping, prehension, writing and the particularly complicated achievements such as tightrope walking or playing the piano, drawing, etc. have to be acquired through practice. This capacity of the organism to learn coordinated movements is maintained in tabes and therapeutic exercise is based upon this fact . . .

. . . It is in the nature of the method that results can be obtained only very slowly. It sometimes takes weeks or months to determine even slight progress. However, and especially in patients who have been bedridden for months and years, one sometimes achieves amazingly fast results on the initiation of the treatment . . .

Disturbances in the Fixation of the Knee and the Pelvis in Neurological Disorders

Verhdlg. Dtsch. Ges. Orthop. Chir. IX, pp. 221–251.

... The most important disturbance in fixation of the pelvis, in my opinion, is the lowering of the pelvis on the side of the circumducted leg in walking, secondary to the failure of the gluteus medius to provide lateral fixation of the supporting leg. I called attention in 1902 to the fact that this tilting of the pelvis and trunk to the side of the circumducted leg represents one of the most important and most constant components of the tabetic gait. Unsteady wobbling gait results. The pelvis and trunk deviate with every step toward the side of the circumducted leg. In order to maintain as far as possible the center of gravity inside the foot base, the circumducted leg is brought down on the ground in a somewhat abducted position resulting in a broad based gait. I am going to show you a picture (Fig. 10 a-4) pertaining to a very severe case of tabes, which shows the complete tiltaway of the pelvis and the pronounced projection of the hip to the right and supporting leg. In other severe cases this pronounced phenomenon may not appear as drastically in walking as it does while standing on one leg. This is particularly true in walking on a narrow base where one leg is placed in front of the other (Fig. 11 a).
... The tightening of the gluteus medius in tabes fails because of the missing sensory stimulation; however, it can be substituted for through voluntary innervation and here lies the key to therapy ...

In the Second World War FOERSTER's great experience contributed to the pre- and postoperative therapy in missile injuries of peripheral nerves. Here he emphasized particularly electro- and exercise therapy.

Directions for the Treatment of Missile Wounds of the Peripheral Nerves

Issued by the Surgeon General, 1939.

... *a) Electrotherapy*
Conservative therapy, which is to be employed first in every case of gunshot injury to a nerve, consists primarily in purposeful and careful *electrotherapy,* that is, the use of galvanic stimulation because it is only through its use that the paralyzed muscle is actually brought to contraction ...

... *c) Passive Movement Exercises*

These are to be carried out several times daily and consist of moving the paralyzed limb back and forth with full utilization of total excursion of the joint, in all its axes ...

Physical Therapy

Handbook of Neurology, Vol. 8: General Neurology. Berlin: Julius Springer 1936, p. 316 ff.

... Most disturbances caused by lesions of the nervous system ... undergo more or less comprehensive compensation even where there is neither reversal of the noxious influence nor regeneration of the destroyed nerve tissue. This occurs by way of a reorganization of the remaining parts of the nervous system, because the latter do not represent a machine composed of individual parts which stop when one part goes out service, but rather they possess an admirable plasticity of amazingly far-reaching adaptability, not only to environmental conditions, but also to functional alterations within the substrate. Occupational therapy facilitates spontaneous restitution, helps it along, builds it up, not infrequently initiates restitution while such powers are lying idle and are not unfolded by the organism, as in the so-called habitual paralyses or in some cerebral hemiplegias ...
... It is not possible to outline a single valid exercise therapy for motor disturbances and diseases of the nervous system. Rather it is necessary to separate analytically the different motor syndromes and to show how the individual components of the syndrome may be compensated for through exercise ...

4. Clinical Syndromes

a) The *Pallidal Syndrome*
b) The *Athetotic Syndrome*
c) The *Choreoathetotic Syndrome*
d) The *Cramp Syndrome*
e) The *Atonic-Astatic Syndrome*
f) *Transection Syndromes of the Spinal Cord*

Of the clinical syndromes in pediatric neurology, one has retained FOERSTER's name. The atonic-astatic syndrome, which has to be considered in the differential diagnosis of myotonia congenita (OPPENHEIM) and in beginning spinal muscular atrophy of the WERNICKE-HOFFMANN type, bears his name. FOERSTER's papers on the more cephalic basal ganglia belong to the classical descriptions of clinical syndromes even though his topographical terms need to be revised today.

On the Analysis and Pathophysiology of Striatal Disturbances of Movement

Zschr. ges. Neurol. and Psychiatr. 73, 1–169 (1921) with 173 pictures.

... As I approach the pathophysiologic explanation of striatal disturbances of movement, I am fully cognizant of the difficulty of this undertaking. I say this after diligent study of the enormous literature and on the basis of intricate investigation of a very large number of pertinent cases, most of them observed for a very long period of time and subjected to a variety of therapeutic methods. The difficulty lies in the great multiplicity of motor types, and the not infrequent admixture of extrastriatal disturbances of motion, and last but not least, in the fact that the anatomical basis for disturbances of motion we consider striatal is as yet insufficiently worked out.

The approach is to ascertain individual, basic types by peeling them out from the manifold patterns which striatal disturbances represent. *They are to be delineated as candidly as possible and analyzed in their basic components in order to attain a pathophysiological explanation on a solid anatomical basis.*

The Pallidal Syndrome: . . . Let us reiterate the individual components of the syndrome of the pallidum. The most important are:

1. Resting tremor, which frequently may be absent.
2. Increase of the plasticity supporting muscle tone.
3. Increase of passive stretch resistance of the muscles (rigor).
4. Development of muscle tension upon passive approximation of their insertion points (adaptation tension, fixation tension, cataleptic behavior).
5. Tonic prolongation of contraction on electrical stimulation.
6. Absence of irradiation of reflex movements, absence of those reflex synergies characteristic of the pyramidal tract syndrome, absence of rebound reaction and tonic prolongation of reflex movements.
7. Absence of reactive movements, absence of expressive movements or perhaps tonic prolongation of these.
8. Reduction of voluntary, spontaneous or initiative movements (paucity of movement), slowing of initiation of movement, slowing of progress of movement, reduced excursion of movement, fatigue and weakening of coarse muscle power in voluntary motions, transient total paralysis in cases of apoplectic origin, tonic prolongation of voluntary movement.

(The pallidal syndrome is further characterized by): . . . absent normal concomitant movements and composite voluntary complex movements, strengthening of normal associated movements, absence of synergistic movements characteristic of the pyramidal system resulting in the maintenance of isolated involuntary movements of single muscles in portions of the limbs . . .

The Athetotic Striatal Syndrome: . . . We have described in summary form the athetotic striatal syndrome in its major characteristics. The following are the main signs:

1. Athetotic play of movements at rest.
2. Diminution between spasms of the plasticity supporting muscle tone.
3. Anomalies of posture of the limbs and trunk corresponding to the crouch position.
4. Over-extensibility of the muscles.
5. Tendency toward fixation of muscle tension, which, however, is inconstant and variable.
6. Inordinantly intensive and extensive movements of reaction and expression with a tendency to tonic prolongation.
7. Pronounced co-innervation and super-imposition of associated movements upon voluntary movement.
8. Inability to sit, stand, or walk. Substitution for these activities by reactive mass movements of the body, reminding one of climbing movements . . .

The Choreoathetotic Syndrome: . . . Chorea is here the prominent feature. The choreoathetotic syndrome consists of the following components:

1. The choreoathetotic play of movement at rest.
2. Diminution of plastic muscle tone.
3. Decreased extensor resistance with over-extensibility of the muscles.
4. Inconstant, fleeting fixation tension of the muscles.
5. Exaggeration of reactive and expressive movements with little tendency toward tonic prolongation.
6. Pronounced co-innervation and excessive associated movements in the performance of voluntary motion.
7. Inability in severe cases to sit, stand or walk, and substitution of these functions by reactive movements of choreoathetotic type . . .

The Striatal Cramp Syndrome: . . . In comparing the cramp syndrome of the striatum, as I would like to term it, with the athetotic syndrome, a strong coincidence of factors becomes evident. The cramping of the muscles of the spine, especially the hyperextension as well as torsions and lateral flexions are not at all unknown in severe general athetosis. They are observed as well in the cramp syndrome. The tendency to spasmodic synergies, the dependency upon affect, emotional or sensory impressions, upon body posture, the aggravation through voluntary innervation are all the same in the two conditions. We find the same over-extensibility of the muscles. However, we cannot speak of a general increase of expressive and reactive movements such as are seen in athetosis. Similarly, there is little typical mass movement as is observed in athetosis upon voluntary motion. In the cramp syndrome, so to speak, the reaction always runs into the synergistic muscle groups that are afflicted with the cramp upon sensory stimulation or emotion. Likewise these muscles are seized with cramp if they or a group of them are innervated voluntarily. Activity, such as standing, sitting or walking, which are severely impaired in athetosis are not affected; however, they may be much impeded by intercurrent cramps and may be altered in their form . . .

The Atonic-Astatic Type of Infantile Cerebral Palsy *

Dtsch. Arch. f. klin. Med. *98,* 216–244 (1910).

. . . In the following presentation I should like to direct attention toward *a fourth type of disturbance of motion in infantile cerebral palsy,* which constitutes a single entity and hitherto has received no special consideration. In some ways this type leans toward cerebellar and perhaps toward choreoathetotic disturbances which, themselves, are closely related . . .

. . . Increased passive mobility is generalized. It is to be attributed primarily to *atonia,* or more correctly *to the lack of involuntary countertension of the muscles* when they are passively stretched. It may be greater than that which we encounter in tabes dorsalis or poliomyelitis with total absence of the respective muscle groups . . .

. . . Secondly, there exists in all our cases an *absolute incompetence of static muscle performance* in addition to the muscular atonia and increased passive mobility . . .

. . . *Ataxia of the trunk when sitting* is as conspicuous as that of the head. Children cannot sit, and according to the position of the gravitational center of the trunk, fall over forward (Figs. 7 and 16) or backward (Fig. 8) or toward the side . . .

. . . Thirdly, *complete ataxia appears in the upright state.* If children are stood upon the floor they collapse like dolls (Fig. 19). The lower legs bend forward over the feet, the knees sink into flexion, the pelvis and upper trunk fall forward over the thighs, the head and spine also curve forward.

. . . Naturally, walking is also impaired. This results because of the utter failure of the supporting leg, which bends in all joints while the circumducted leg is well fixed at foot, knee, and hip and is brought forward . . .

. . . The total inability to steady the head, to sit, to stand, as existed in all our cases, made us think at first glance of a complete paralysis of the respective muscles. This is not the

* Later named after FOERSTER.

case at all. On the contrary, during the period of a few weeks postpartum, the children, according to the mother, perform no motion but rather lie with the head, trunk and extremities immobile. Following that period, however, all parts of the body are moved vigorously . . .

. . . On the one hand the muscles are not atrophic and demonstrate no change in electrical excitability. On the other hand, *the deep tendon reflexes are preserved . . .*

. . . Secondly, there exists in our cases a typical disturbance of coordination manifesting itself in vigorous action of individual muscle groups, if they are called on as synergists. In other words, all individual motions may be carried out by the muscles, but they *fail totally in their static functions* (fixation of the head, fixation of the trunk and the spine in sitting, fixation of segments of the leg and trunk in standing and walking) and they likewise fail if, as *antagonists,* they should have *to dampen a motion* or even arrest it in a moment; or if, as *collateral synergists,* they should have to fix a moving limb in the plane of the motion . . .

The Syndromes Seen in Transections of the Spinal Cord

Handbook of Neurology, Vol. 5: General Symptomatology III, p. 106, 1936.

Even today there is a no more exact description of the so-called transverse syndromes resulting from cord transection than that provided by FOERSTER in the Handbook of Neurology.

Cord Transection at C7

. . . After transection of the medulla in the area of the 7th cervical segment, supraspinatus, infraspinatus and teres minor, deltoid, biceps, brachialis, brachioradialis and supinator brevis remain completely intact. In addition the function of subscapularis, pectoralis major, latissimus, teres major and pronator teres are retained, at least partially. Also, I found the function of the extensor carpi radialis longus relatively well preserved in all pertinent cases. However, I found the triceps always totally paralyzed. I emphasize this in contradistinction to KOCHER, who lists the triceps among the non-paralyzed muscles in the transverse syndrome of C7. Aside from the extensor carpi radialis longus, all hand and finger muscles are paralyzed. The small influx of innervation which the flexor carpi radialis and the extensor digitorum communis not infrequently derive from C6 is insufficient in a transection in the area of C7 to insure a capacity of these muscles to tip the scales to any degree. The resting position of the upper extremity in the transverse syndrome, C7, is characterized chiefly by the preponderance of the forearm flexors over the paralyzed extensors; the forearms are held flexed in an acute angle (Figure 76). The high position of the shoulder, which is so characteristic for the transverse syndrome of C6, is much less evident in the transverse syndrome C7, because the shoulder depressors (pectoralis and latissimus) are largely spared from paralysis. Because of the relative integrity of the internal rotators of the humerus (Subscapularis, pectoralis major and latissimus) the external rotation of the humerus is absent—it is not infrequently present in the transverse syndrome C6. In the same fashion, diminished activity of the pectoralis, latissimus and teres major, which function as abductors of the upper arm, makes the abducted position of the humerus less evident in transverse lesions at C6. Even the supinated position of the hand, so pronounced in the transverse syndrome C6, may be much less evident in the transverse syndrome C7, because of the relative integrity of the pronator teres . . .

5. Neuroradiological Experiences

FOERSTER promptly accepted DANDY's ventriculography (according to BINGEL's method) and he reported in 1925 his experiences in one hundred cases of encephalography. He laid particular emphasis upon the early acquired cerebral damage, the findings in hemiplegia in the adult, epilepsy, tumors (such as a quadrigeminal plate tumor) and pseudotumor cerebri, as well as the findings in encephalitis and trauma. In this last field, FOERSTER had unusually wide experience.

Encephalographic Experiences

Zschr. ges. Neurol. Psychiat. *94*, 512–584 (1925).

. . . The encephalographic findings in traumatic brain lesions show again that the ventricle is an extraordinarily fine test object for the process in the hemisphere produced by the trauma. We find enlargement of the ventricle corresponding to the side of the injury even where no real focal symptoms are manifested clinically (case 39), or where such symptoms appear transiently only as in case 41. One notices particularly well in lateral pictures the direct relation of the enlargement to the site of the brain lesion (case 40). The ventricle appears almost attracted by the lesion. Earlier in numerous gun-shot wounds of the brain, I convinced myself over and over again by ventricular puncture that a high degree of enlargement of the ventricle is present in the injured hemisphere. Unilateral, internal hydrocephalus is almost always part of the picture of gun-shot wound to the cerebral hemisphere. In addition, I have repeatedly observed the close correlation between ventricular enlargement and the site of the lesion. I stated previously that in the lateral encephalogram, the ventricle appears as though attracted toward the site of the lesion (Case 40). I look on this as an actual "migration of the ventricle" toward the lesion. This occurs slowly and gradually . . .

II. Contributions to Neurosurgery

1. Posterior Root Section for the Diminution of Spasticity ("FOERSTER'S Operation")

FOERSTER *Introduces Posterior Root Section*

His experience with physiotherapy in tabetics and the old observation that hemiplegias in tabetics remained flaccid in the area affected by the tabetic process gave FOERSTER the idea that operative destruction of the posterior roots might eliminate the cerebrally conditioned spasticity. Thus in 1908 he suggested posterior root section (an intervention later termed FOERSTER's operation), which rapidly made his name internationally known. However, it has not held the promise originally expected by FOERSTER, particularly since the cause of this "cerebral" spasticity is no better understood today, despite the utilization of the "gamma system" concept as an explanation of its pathogenesis. On the other hand the operation of posterior root section achieves much more in pure spinal spasticity characterized by the preponderance of so-called flexor-reflex synergies.

On a New Operative Method of Treating Spastic Paralyses by Resection of Posterior Spinal Roots.

Zschr. Orthop. Chir. 22, 203—223 (1908).

... Gentlemen: We have only to follow this hint of human pathology. If the spastic muscle contracture is truly based upon a reflex acting without inhibition because of loss of pyramidal tract control, then we should be able *to cancel it out by operatively transecting one link in the chain of this reflex arc.* The motor components of the reflex arc, that is, anterior horn, anterior root, motor nerve, naturally cannot be sacrificed, since although their elimination might well eliminate the contracture, it would, at the same time, produce a complete and flaccid paralysis of the contracted muscles. On the sensory side of the reflex arc, the peripheral sensory nerves are mixed everywhere so intimately with the motor fibers, that their isolated elimination is impossible. The experiment executed by nature, that is the interruption of the posterior columns at the entrance zone of the roots, cannot be taken into consideration either since the spinal cord represents a *noli me tangere.* The posterior root remains as the only part of the sensory portion of the reflex arc subject to interruption ...

... As we now want to put this general thesis to a practical test, we have to state in a given case which muscle groups are predominantly in contracture and which spinal segments mediate the reflex irritability of these muscle groups. One makes his choice of these segments, preferably in such a way as not to remove two roots in sequence.

... However, the advantage offered the boy by operation is by no means exhausted in the *disappearance of the contracture and the restitution of an approximately normal passive mobility and range of excursion of the limbs.* As you see, active mobility has returned to a great degree as well. It is particularly noteworthy that isolated movements of the leg and of individual parts of the leg may be carried out voluntarily. The foot can be fully dorsi and plantar flexed without flexion or extension occurring respectively in knee or hip ...

... In the case just described, we dealt with a spastic paraplegia of cortical origin. In this second case we have a spastic paraplegia of spinal origin ... a ten year old girl, who has suffered from *tuberculous spondylitis of the cervical spine* since the third year of life. Therefore, there exist marked gibbus, progressive compression of the spinal cord and paralysis of the legs. She has been unable to walk since Christmas 1905 ... Walking is impossible, even with support, since the patient cannot move the right leg at all and is barely able, at times, to move the left ... Since operation, voluntary mobility has increased significantly too, particularly on the left ... The involuntary spasmodic flexor movements of the legs, which were so bothersome to the youngster before operation, have disappeared entirely. Neither do the legs develop spasmodic flexion as associated movements when the child sits up.

... One short word on the indications for surgery. In my opinion, this method is suitable for all severe spastic paraplegias of the legs, no matter whether based upon cortical or spinal disease ...

2. Posterior Root Section in Gastric Crises

The careful study of so many tabetics during physiotherapy had acquainted him with the tortures of "lancinating pains" and of "gastric crises". He had had the occasion to study the main pain pathways in the spinal cord after posterior root section. Thus the idea occurred to him to eliminate these painful syndromes by resection of posterior roots.

On the Operative Treatment of Gastric Crises by Resection of the 7th—10th Posterior roots

Beitr. klin. Chir. *63*, 245—256 (1909).

... Because we start from the idea that the basis for gastric crises is a pathological irritation of the sensory fibers of the stomach (and it represents in its entirety a

phenomenon of sensory irritation) we must ask first which *nerves serve as the sensory supply of the stomach.* Physiology teaches that the vagus contains gastric sensory fibers. But a part of the sensory supply certainly passes through sympathetic fibers which pass from the stomach into the celiac plexus and from there into the major splanchnic nerves, thence by rami communicantes into the dorsal roots of the spinal cord. HEAD's investigations show it is particularly the 7th—9th thoracic posterior roots that conduct the sensory sympathetic fibers of the stomach . . .
. . . Since on one hand all therapy hitherto had been a failure, and on the other hand the condition represented unspeakable suffering and a great danger to the patient's life (he had starved down to a skeleton and had literally lived on morphine for weeks) we decided in this instance to interrupt the *pathways carrying the pain impulses of the gastric crises.* According to our thesis, *gastric crises are to be viewed as a pathological state of irritation of the 7th to 9th posterior dorsal roots. Resection of these roots was performed . . .*

FOERSTER reports on the results of posterior root section at the International Congress for Internal Medicine in London.

Resection of the Posterior Spinal Nerve-Roots in the Treatment of Gastric Crises and Spastic Paralysis

Proc. Royal Soc. Med., July 1911.

. . . Last summer, GOTTSTEIN and I cut the fourth and fifth lumbar and first sacral roots in a tabetic patient suffering from a perfectly localized neuralgia of the internal malleolus; for some weeks there was complete absence of pain, but afterwards it recurred as badly as before. I am inclined to believe that if resection of the posterior nerve-roots is to be performed at all in cases of tabes with lightning pains, it will be necessary to remove all the roots of one extremity at once; for we never know with absolute certainty which roots are particularly influenced by the morbid stimuli, and there is considerable overlapping of the regions supplied by the different roots . . .
. . . When I suggested performing resection of certain posterior thoracic roots in cases of visceral crises, I was influenced by a consideration of the severe pain and symptoms of sensory irritation which form the basis of these crises, and of the often enormously increased hyperaesthesia of the skin of the epi-, meso- and hypogastrium which accompany the crisis . . .
. . . The operation has now been performed altogether twenty-eight times, as shown in Table I (p. 228). Three cases succumbed to the immediate effects of the operation, two

showed no improvement, the crises persisting. In the remaining twenty-three cases the operation was successful. Immediately after the resection of the roots the crises disappeared, the body-weight increased, and the general condition showed a marked improvement: some of the patients who, until then, had been absolutely confined to their bed for some time, have since regained their power to work. In the majority of these cases (fifteen) no relapse has been reported ...

... I will now discuss the value of resection of the posterior spinal nerve-roots in spastic paralysis due to disease of the cortico-spinal path, especially the pyramidal tract ...

... The operation has hitherto been performed in eighty-one cases, as is shown in Table II. Of these, nine died as the result of the operation; seventy-two survived; fifty-one were cases of congenital spastic paraplegia (LITTLE's disease); almost all of these patients have been benefitted by the operation, some of them showing a marked improvement ...

3. Section of the Pain Tract in the Spinal Cord, the So-called Antero-Lateral Tract Section

FOERSTER *takes the lead in the surgery of pain*

Posterior root section in the relief of lancinating pains and gastric crises constituted the first step in pain surgery. A complication in one of his own cases of posterior root section which led to unbearable pain caused him to try in 1912, together with the general surgeon TIETZE, transection of the anterolateral funiculus–independent of SPILLER and MARTIN, even though one year later. This method remains to this day one of the most important procedures for the elimination of pain. The extent of the transection of the anterolateral funiculus was later carefully examined by him and GAGEL. Their results constitute a basic pathophysiologic and morphologic study of the spino-thalamic tract. Whether the temperature fibers actually lie as far dorsal as was assumed by FOERSTER, and whether the groupings of pain fibers are really laminated like onion peel, are still matters of discussion today. However, his study is without any doubt the starting point for all scientific investigations of cordotomy.

Anterolateral Funiculus Section in the Spinal Cord for Elimination of Pain

Berl. klin. Wschr. *50*, 1499–1502 (1913).

... We intended to resect also the 1st and 2nd lumbar roots, which participate in the sensory supply of the lower abdominal area. We went in through the lower corner of the incision and ran into necrotic, putrid material in the lower part of the old operative site, which may well have been the cause of the pain. This material was scraped out and in doing so, the dura was torn in one place. There was no discharge of spinal fluid, however, and the subarachnoid space was not opened because a thin layer of arachnoid had been left intact. We were afraid an infection of the meninges might ensue through this thin layer and we, therefore, painted some tincture of iodine upon the dural tear with the intent of causing adhesions as speedily as possible. There was no infection, but great misery resulted from the iodinization. Obviously there was an intense irritation of the arachnoid and the result was an altogether intolerable pain in the entire left leg. The iodine application had been about at the level of the 12th thoracid vertebra, that is at the level of the lumbar enlargement. An increasingly rapid atrophy and paresis of the left leg followed. The pain was of intense character, lasted day and night, and could only be helped somewhat by large doses of morphine. The patient categorically demanded to be helped out of this intolerable situation through further surgery. I considered *re-exploration of the spinal cord at the operative site* and a section of all left lumbosacral roots at the lumbar enlargement, a procedure that would be technically difficult to carry out because of the suspected adhesions. Therefore, I suggested a transection of the crossed right anterolateral funiculus of the spinal cord, which carries the pain fibers for the left leg. This operation was carried out in the upper thoracic cord by Mr. TIETZE and me in December, 1912. I approached the anterolateral funiculus, turning the *spinal cord somewhat toward me by means of traction on the anterior root. Just anterior to the dentate ligament I simply inserted a very fine knife, cut the cord anteriorly and medially so that the knife point would emerge again lateral to the anterior spinal artery. The insertion must be carried out anterior to the ligament so that the pyramidal tract is not injured. Of course, the anterior spinal artery must be spared lest paraplegia result.* The success was stupendous. The pain in the left leg was eliminated with the keenness of an experiment. There was analgesia of the left half of the body up to the height of the left nipple. The sense of touch was in no way influenced by the operation. There was not the slightest paresis in the right leg, not even a Babinski could be manifested in either leg. There was transient bladder weakness, but this disappeared rapidly. The pain in the left leg has been permanently abolished . . .

Anterolateral Funiculus Section in Man.
A Clinico-patho-physiologic-anatomical Study
With O. GAGEL. Zschr. ges. Neurol. u. Psychiatr. 7, 138, 1–92 (1932).

. . . We now carry out section of the anterolateral funiculus by inserting the point of the blade into the cord immediately in front of the dentate ligament. It has happened several times that, in the effort not to injure the immediately and posteriorly placed posterolateral funiculus (which carries the pyramidal tract), we initially introduced the blade too far ventrally and thus spared the dorsal segments of the anterolateral funiculus. In these cases there was no initial disturbance of temperature perception, whereas pain perception was eliminated entirely. As we reintroduced the instrument and cut the dorsal portions of the anterolateral funiculus as well there was added disturbance of temperature perception.

... The other question concerning the organization of the anterolateral funiculus is whether the fibers stemming from the individual posterior horn segments mix in their ascending intrinsic course or wheter they retain a position separated from each other; in other words, whether the anterolateral funiculus has a segmental organization. In our opinion, this has to be affirmed unconditionally, as we have indicated in previous papers. We are of the opinion (Fig. 53) that the tracts from the individual posterior horn segments ascending in the anterolateral funiculus are grouped according to the *same principle of onion peel organization* as are the other long tracts of the spinal cord. They are grouped in concentric half circles around the grey matter of the anterior horn in such a fashion that the outermost layers contain the tracts from the most caudal segments, the innermost lying layers contain the fibers from the most cephalic posterior horn segments. Every fiber contingent stemming from a posterior horn segment attaches itself from the inside to the fibers already present from lower segments as they enter into the crossed anterolateral funiculus. This concept of the segmental organization of the anterolateral funiculus has been confirmed fully by our experience with cordotomies ...

... H. HEAD had already assumed a layered segmental organization of the anterolateral funiculus, such as we have detailed here. He has reported cases of disease of the anterolateral funiculus in whom the lower sacral dermatomes S5 to S3 or even S5 to S1 had been spared from sensory disturbance (Figs. 57 and 58). We too have reported similar cases on other occasions ...

4. Spinal Cord Tumors

FOERSTER *Operates on Spinal Cord Tumors*

During and after the First World War FOERSTER became a surgeon of the spinal cord. As early as 1917 he was able to report on the first successful removal of an intramedullary tumor. In 1920 he reported the removal of 12 spinal cord tumors, of which nine showed significant restitution of function. In a paper with BAILEY appearing in 1936 in the Jubilee Volume for DAVIDENKOW, he reported on his experiences in this field in detail, recommending certain techniques and comparing their value with that of other procedures. In the interval the technique of this operation had advanced further and many medullary tumors had been operated radically. FOERSTER, himself, added to his own experience, and he reported in 1935 on the operative treatment of 88 spinal cord tumors of which 30 were extradural, 33 were intradural and extramedullary and 20 were intramedullary. Of these, seven died as a consequence of the operation.

A Contribution to the Study of Gliomas of the Spinal Cord with Special Reference to their Operability

With PERCIVAL BAILEY, State Institute for Public Biology and Medicine, Literature "Jubilee Volume for Davidenkow", pp. 6–67, 1936.

. . . When a glioma of the spinal cord is disclosed at operation, many procedures are possible, which one may evaluate in the light of the pathology as follows:

1. One may close the dura mater without disturbing the tumor. The result in the one instance recorded where such a procedure was followed (DIVRY and TECQUEMEME) does not incline to this method. The tumor is often under tension, which would make the closure of the dura mater difficult.

2. One may leave the dura mater open without disturbing the tumor. Prompt amelioration of symptoms may follow (BAILEY and BUCY) and, if the tumor is of benign type, may last for months or years.

3. One may split the dorsal columns and leave the dura mater open. This is the favorite approach to these tumors. Their predominant situation in the dorsal columns justifies it, as will the fact that it is the easiest and most direct route to follow. For infiltrative tumors one should certainly be content with decompressing the tumor by splitting the pia mater and overlying cord.

4. After proceedings as in 3. one may later reopen the wound and attempt to extirpate the tumor, which may in the meantime have partially extruded itself from the cord. This is the method of ELSBERG and BEER and in our experience has not been very successful.

5. One may, after proceedings as in 2., subject the patient to roentgenradiation . . .

6. Finally one may attempt to dissect out the tumor at the first operation. Our own experience and a study of the literature makes us inclined to think this procedure rarely advisable.

5. Surgery of Peripheral Nerves

FOERSTER, Himself, Undertakes the Surgery of Peripheral Nerves During the First World War

The great number of gunshot wounds of the peripheral nerves in the First World War led FOERSTER to enlarge the experience he had gained from working with TIETZE and KÜTTNER during the prewar period, undertaking the surgery himself. During the war he was able to report on a large number of operations on peripheral nerves (nerve suture, neurolysis) at the meeting of military surgeons in 1917, where he was able to summarize 4, 748 observations of his own. Of these he had operated personally upon 775 patients. He reported upon his large experience in the three supplementary volumes of the First Handbook of Neurology. In the first volume he provided an anatomy of the peripheral nerves, which originated from a gargantuan study of the literature and from his own extensive experience. Superb pictures illustrated the work. In the second volume he described the muscles and their functions. Finally, in the third volume, he described the surgery of the nerves on the basis of his operative experience. FOERSTER put particular emphasis upon pre- and postoperative therapy with such electric currents as would bring the muscles to contract (cf. p. 37). Because of the absence of antibiotics and the increased danger of infection in primary operations, he was a strict partisan of secondary suture, the optimal time for interference being between four and six months, depending upon the length of the nerve.

The Operative Treatment of Gunshot Wounds of Peripheral Nerves

Münch. med. Wschr. 31, 1183 (1934).

... It becomes evident from the details so summarily presented above that a great number of the gunshot wounds of peripheral nerves, which I observed during the World War, were capable of spontaneous repair. Of the 4,748 gunshot wounds to nerves which I observed during the World War, 1,018 cases are excluded from consideration of the rate of spontaneous restitution versus those not amenable to spontaneous restitution because they were not followed by me for a sufficient period of time. Also, I eliminate from the *statistical consideration* all those cases—815 in number—which were lesions of pure sensory nerves because they have to be judged altogether differently from the cases of lesions of motor or mixed nerve trunks respectively. There remain for consideration 2,915 cases, whose courses could be followed accurately. Among those, 1,320 cases, that is 45% of all cases, came to a spontaneous cure, 660 cases, that is 22%, came to more or less considerable improvement, 955 cases, that is 33% came to no or a practically insignificant spontaneous restitution so that an operative intervention appeared indicated as far as the nerve injury was concerned. Of these 955 spontaneously irreparable cases operation was declined in 180 cases (6.5%) or it could not be carried out because of particular complications (pseudarthroses, extensive cicatrix formation, fistula formation, etc.); the other 775 cases (26.5%) I have operated upon. That means there was reason for surgery or possibility thereof in only 26.5% of the cases . . .

Without going into details, it may be said that in 370 cases of pure nerve suture there occurred a cure in 55%, an improvement in 42% and no success in only 3%. Of course, one would have to break down such statistics according to the nerve, the site and the type of injury in order to obtain a clear picture of operability.

6. Brain Surgery: Elimination of the Meningo-cerebral Scar

The Operative Management of the Meningo-cerebral Cicatrix in Patients Developing Seizures Secondary to Gunshot Wounds of the Brain

The injured of the First World War, who developed convulsive attacks after gunshot wounds of the brain posed a real problem: that of the elimination of the attacks. FOERSTER thought that as a result of the incarceration of vessels in scar tissue, angiospastic phenomena developed which, in turn, led to convulsive seizures. The operative lysis of the cicatrical area was a logical sequel. With PENFIELD, he reported on the initial results in a comprehensive paper. Here he gave not only a synopsis of the good operative results, but also, as a "byproduct", highly interesting insights into the results of cortical stimulation during these operations. He employed stimulation to provide a more accurate physiological analysis of the scar area, in order to "orient himself accurately". The result was a finer "localization of functions" on the "brain map".

The Structural Basis of Traumatic Epilepsy and Results of Radical Operation

With W. PENFIELD. Brain 53, 99–119 (1930).

. . . Exploration of the human cerebral cortex, both in normal and pathological areas in well over 100 cases under local anesthesia, has made it possible to outline certain definite epileptogenic cortical areas (FOERSTER). The areas in the human cortex are analogous to but not identical with the areas outlined by VOGT in monkeys. The movement patterns which follow such stimulation experiments provide a local sign, which often makes it possible to localize the irritative focus of an epileptic discharge, even where there are no physical signs to suggest this localization . . .

. . . We have shown that at operation the focal epileptic attacks may often be produced in two ways: either by electrical stimulation of the brain in the neighborhood of the wound, or by gently pulling upon the adherent dura. This latter fact may be of considerable significance, for if increase of a pre-existing strain produces an attack it may well be that preexisting strain itself is an important factor in the aetiology of spontaneous convulsions.

As pointed out above, the blood-vessels form in one sense the woof of the contracting network. Traction, therefore, upon the vessels must be inevitable. The hypothesis at once suggests itself that a vasomotor reflex secondary to this traction is responsible for the initiation of the convulsive seizures . . .

7. Operative Treatment of Brain Tumors

FOERSTER *Embarks on the Surgery of Brain Tumors. He Performs the Second Successful Removal of a Tumor of the Quadrigeminal Plate!*

Experience with the surgery of meningo-cerebral scars led FOERSTER to tackle further problems of neurosurgery, especially the operative removal of brain tumors. He was second only to FEDOR KRAUSE (1913) in succesfully removing, in 1928 a large tumor of the quadrigeminal plate. He reported on his experiences with brain tumors in 1934. A comprehensive paper dealing with the subject, which was presented to the British Association of Neurological Surgeons when that Society visited him in Breslau in June of 1937, has never been published. His call for the early diagnosis of brain tumor by the general practitioner is as valid today as it was then.

A Case of Tumor of the Quadrigeminal Plate Removed by Operation

Arch. Psychiat. *84*, 515–516 (1928).

. . . Diagnosis: Tumor of the quadrigeminal area. Ventriculography shows greatly dilated lateral ventricles and a dilated third ventricle. Operation: Exposure of the right occipital lobe to the transverse sinus and longitudinal sinus. Ligation of all veins leading from the cerebrum into the latter. Along the falx, headway is made towards the tentorium and the splenium of the corpus callosum. Split tentorium anteroposteriorly along the sinus rectus. Division of splenium. This brings the tumor into view; it is exposed as much as possible on all sides and it has the size of a tangerine. It is resected piecemeal and completely; bleeding slight. After its removal, the vein of Galen comes into sight, displaced to the left. Transient respiratory paralysis is corrected promptly with Lobelin. Nature of tumor: Glioma. Healing per primam. Complete resolution of all symptoms except for blindness; even now one half year after surgery there is light perception only. Pupils react slowly but strongly. Hearing normal, ocular movements show no restriction. Slight nystagmoid jerkings. No disturbance of equilibrium, no ataxia, no paresis, no sensory disturbances, no spontaneous choreiform movements.

The Diagnosis and Treatment of Tumor of the Cerebrum

Klin. Wschr. *13*, 1737–1742 (1934).

... Two tasks delineate themselves very clearly for the future. The first concerns the diagnosis of tumors. Everything depends upon a diagnosis as early as possible. However, the entire medical profession has to cooperate. Every doctor must know in which case he has to consider the possibility of a brain tumor. This is altogether sufficient; he needs only to direct the case in question to the neurologist for further clarification. The second task concerns therapy. Concerning operative indications, it will be our intent in the future to separate the wheat from the chaff. There is no purpose in operating on hopeless cases. Even if in a malignant glioblastoma or in a medulloblastoma the lifespan of the individual in question is prolonged through operation, or even if a final and truly radical elimination of the tumor should be possible through a very radical approach, such as through extirpation of an entire hemisphere (and to decide on that, our past observation time is not sufficient) these patients always remain pitiable cripples for the rest of their lives ... It is a paramount need therefore to recognize where possible in advance which type of tumor is present in the individual case, and to determine whether operative intervention is promising. Unfortunately, we are not today as advanced as that, but I am convinced that we shall get there by deeper penetration into the symptomatology and its development in each individual case, and by utilization of all possible diagnostic methods on the one hand, and intense histological study of tumor types on the other hand ...

8. Operations on the Vegetative System

Foerster also had great interest in operations on the vegetative system. His monograph on pain sensation and its pathways of conduction (1927) contains many examples of successful intervention upon vegetative structures. It is evident from a letter of Foerster to Kahn that it was probably Foerster who carried out for the first time the supradiaphragmatic resection of the splanchnic nerves, an operation which later played a role in the treatment of hypertension under the name of Adson's procedure.

In 1939 Foerster reported comprehensively on his observations during operation on the vegetative system, their scientific implications, and their relation to an understanding of the circulation. In that paper he describes the effect of spinal anterolateral quadrant section upon blood pressure. An assay of this suggestion, using significantly larger number of patients, was not pursued.

Operative-Experimental Experiences in Man on the Influence of the Nervous System upon the Circulation

Verh. Ges. Deutsch. Neur. Psychiat., 5. Jahresvers. Wiesbaden. Z. ges. Neurol. Psychiat. *167*, 456 (1939).

. . . Observations in the human which we have collected in approximately 100 cases of anterolateral quadrant section have given us information concerning the position of this *diencephalic vasomotor tract* in the transverse section of the spinal cord. Very soon after I had started to carry out section of the anterolateral funiculus—I did this for the first time in 1912, that is 27 years ago—it struck me that in some cases the blood pressure dropped considerably immediately thereafter, in several instances even to a menacing degree; to recover only very slowly and partially. Our special attention was not aroused until, in several cases who previously had been suffering from an advanced grade of hypertension, the high pressure disappeared after the section of the anterolateral funiculus. In one case the blood pressure dropped from 270 to 100, and even though it subsequently recovered some, it never again exceeded the value of 140.

. . . Accurate survey now showed that a substantial drop in blood pressure only ensues where the cordotomy knife takes in the dorsal-most fiber layers of the anterolateral funiculus, so that the conclusion is justified that the supranuclear vasoconstrictor tract descending in the spinal cord (Fig. 18) is situated predominantly in the posterior part of the anterolateral funiculus, immediately ventral to the lateral pyramidal tract . . .

III. Contributions to the Applied Physiology of the Nervous System

1. The Dermatomes

The functional analysis of movement had demanded as a prerequisite an accurate study of the reflex arc and of the motor tracts. Later on, Foerster utilized every operation to determine how the operative transection affected neural function, for example, the reflex arc. Also, sensory defects were stated exactly. Soon he supplemented this "negative" experiment by the "positive", the electrical stimulation of the substrate during surgery; and in this instance, initially the anterior and posterior roots. Thus exact maps of the borders of the dermatomes were developed, which he reported in the Schorstein Memorial Lecture. He analyzed them by three methods: 1. determination of the upper and the lower limits after transection of the posterior roots. The number of his cases of posterior root section offered to him the occasion to determine practically every single root. 2) If he sectioned several roots at a time, he usually cut the uppermost and lowermost roots first to determine the extent of any root isolated between, which otherwise was not possible because of the overlap of the dermatome borders. Besides this determination of the upper and lower root border and that of the isolated root he observed; 3) that by faradic stimulation of the "vasodilators" the supply area of the stimulated root could be rendered recognizable, because its entire tributary area became strongly hyperemic. Thus he had a triple check on his results. Stimulation of the anterior and posterior roots and corresponding transection, furthermore, offered him the possibility to study satisfactorily the peripheral innervation and check it electrophysiologically. He also determined the role in sensation of the anterior root.

Symptomatology of the Diseases of the Spinal Cord and its Roots

Handbook of Neurology, Vol. V, by O. Bumke and O. Foerster. Berlin: Julius Springer Press 1936.

... So far our knowledge of the dermatomes in the human body is based partly upon the anatomical investigation of Bolk, already cited, but particularly upon the fundamental studies of Henry Head, who established the first well founded schema of human dermatomes (Fig. 149 a, b, c, d) ...

... In thirty years of neurosurgical endeavor I have had occasion to determine in man a very large number of dermatomes employing Sherrington's method of residual sensation. These were cases of severe spastic leg or arm paralysis in whom resection of the posterior lumbosacral or cervical roots was carried out to eliminate spastic contractures; thereby one single posterior root had been left intact so that its respective dermatome signals clearly the zone of residual sensibility. Corresponding observations for thoracic dermatomes are to date lacking. However, we may utilize for the determination of the borders of human dermatomes all cases in whom a number of neighboring roots have been transected because it is quite evident that in such a case the oral border of the respective resulting anesthetic zone represents the caudal border of the next higher dermatome, and conversely, the caudal border of the anesthetic area represents the oral border of the next lower dermatome. Because I have carried out section of several roots at all levels of the spinal cord and therefore have had the opportunity to select the most varied combinations, the oral and caudal borders of almost all dermatomes in the human could be delineated, and we are, therefore, in a position *to construct* the site, shape, and extent of almost every dermatome. Fig. 150 and 151 render the dermatome schema of the human as attained by me through the method of residual sensation and the method of construction . . .

. . . So far we do not have dermatome determinations in man by means of strychnine intoxication of posterior roots, the method of DUSSER de BARENNE. However, we may utilize in our analysis certain cases from human pathology where an irritative agent influences one single posterior root and produces a selective oversensibility of the corresponding dermatome. Fig. 152 shows the sharply delineated hyperesthetic skin zone corresponding to the 6th thoracic dermatome in a case of a root neurinoma growing from the 6th posterior thoracic root (Fig. 153). In this case the determination of the root involved could be based exclusively upon the sharply limited hyperesthetic zone and the tumor was removed operatively. Such more or less sharply circumscribed dermatomal hyperesthesias occur in other irritative root processes too, particularly at the onset of herpes zoster, in syphilitic radiculitis, traumatic root irritation, etc. . . .

. . . The only posterior root resection after which there results complete de-afferentation of a certain skin area is the 2nd cervical root. According to my investigations so far, neither the 3rd cervical dermatome nor the tributary area of the cranial nerves (trigeminal, intermedius, vagus) extended significantly over the occiput; therefore, there is complete loss of skin sensation in that area after isolated resection of C2 . . . I have noticed that particularly in the trunk, the vasodilator dermatomes occasionally cross the anterior midline by 1—2 cm . . .

2. The Sensation of Pain and its Conduction Pathways

Later, FOERSTER reported in a monograph his work on the physiology and surgery of pain. He based his study upon his own tremendous experience. He described an accessory pain tract in the anterior root and another in the adventitial plexus of the great arteries. He considered the significance of the postcentral gryrus in the origin of pain. Even the old subject "Soul and Pain" was broached.

The Conduction Pathways of Pain

Berlin and Wien: Urban and Schwarzenberg Press 1927.

... If, for therapeutic purposes we carry out interruption of a nerve, for example by means of intraneural novocaine or alcohol infiltration or by means of ethylene chloride freezing we find that of all the fibers contained in the nerve trunk the pain fibers yield the last ...

... In my material comprising about one hundred root sections there is only one case, a tabetic, in whom I severed three adjacent roots, L4, L5, S1, and could not manifest the slightest sensory defect in the skin ... This observation speaks decidely for the existence of an accessory tract for the conduction of common sensory and pain impulses other than the posterior root which, should there be interruption of the main pathway formed by the posterior roots, may substitute to a varying degree and obviously to a varying extent...

(see the following paper by FOERSTER p. 61)

... In 1924 I formulated my thesis as follows: the posterior as well as the anterior roots carry afferent pathways. The posterior roots represent the main sensory system; their elimination always produces sensory defects if a sufficient number of roots are severed. The anterior roots represent a mere auxiliary pathway for sensation; their isolated interruption is never followed by tangible sensory defects. However, in an interruption of the main pathway (the posterior roots) they may cover or compensate to a varying degree for the sensory defect caused by destruction of the posterior roots. The vicarious function of the anterior roots relates mainly to deep sensibility. However, cutaneous nerve fibers too are likely to pass through the anterior roots.

... The strict proof that the anterior roots in the human actually contain afferents, that is pain conducting fibers, has also been offered by me, for I have demonstrated on numerous occasions that severe pain originates from electrical stimulation of the central stump of a transected anterior root. It is localized accurately by the patient to the segmental zone corresponding to the respective root ...

... Deep excision of larger segments of the upper parietal lobe are followed by disturbance of sensation of the entire contralateral half of the body ...

. . . If any disease process acts as an irritative agent upon a circumscribed area of the postcentral gyrus, it is known to lead to paroxysmally appearing irritative sensory manifestations, generally more or less strong paresthesias, which at times may even have a painful character. These paresthesias start in that part of the body whose focus in the postcentral gyrus constitutes the point of attack of the irritative agent; these paresthesias propagate successively, like a wave over the entire contralateral half of the body . . . I have never succeeded in eliciting any paresthesias, even with the strongest stimulation, in any portion of the cortex other than the postcentral gyrus and the upper parietal lobe . . . The question concerning us immediately in this paper, and which we have stated at the beginning of this chapter, is whether the sensation of pain is tied to the activity of the cortex, and if its collaboration is necessary at all for the production of it . . . I believe it justifiable to conclude that the pain and paresthesia projected into the extremities during cortical stimulation actually represent the psychic equivalent of excitation of cortical sensory elements . . .

. . . If we try to reduce the mode of action of the manifold methods of psychotherapy to a common basis there are, in my opinion, two factors that are more or less important in all methods. Psychogenic pain is produced and sustained in the majority of cases by affects of distaste, anxiety, fright, fear, anger, tension, distastefully accentuated expectation. Conversely, psychogenic pains is cancelled out through contrary affects, pleasure, hope, happiness, joyful expectation and similar emotional reactions. We have to recognize the dependence of pain upon the affect as an almost generally valid principle. Anyone subject to swings of mood will corroborate this . . .

. . . In addition to this first factor, psychotherapeutic methods operate to change affect. There comes, in my opinion, at least in some cases, another factor which has to do with the activity of the will. The question how far the will is in the position to influence pain is not very easily answered, even though at first glance it may appear so and it is taken for granted by most people . . . The question of influence upon pain by the action of intent leads to the question of influence of attention upon pain . . . Finally, I should like to recall a special example clearly demonstrating the importance of distraction of attention and shedding unique light upon the subject. It has been known for a long time and has been documented a thousandfold over during the First World War that through concentration of the psyche upon the action of battle sometimes even the most severe injuries do not reach the conscious level. The psychological analysis of such relationships is exceedingly difficult and the attempt to reduce the process to a neurodynamic correlate could only succeed by chance.

Foerster also compiled (1927) the following characteristics of "hyperpathic pain". Hyperpathic pain originates

1. In the face of a relatively high pain threshold and as an inadequate reaction to the stimulus.
2. After prolonged duration of the sensation, perhaps with summation.
3. With latency of pain sensation.
4. With explosive eruption.
5. With abnormally disagreeable character.
6. With prolongation and eventually with a pain free interval.
7. With deficient localization of the stimulus.
8. With irradiation and
9. with lack of special coloring of the sensation.

On the Relations of the Vegetative Nervous System to Sensation

With H. Altenburger and F. Kroll. Zschr. ges. Neurol. Psychiat. 121, 139–185 (1929).

... Not long ago we observed a patient who in the war had suffered a gunshot wound of the tibial nerve; there was complete paralysis in the distribution of the nerve, and at the same time a constantly recurring trophic ulcer on the sole of the foot, resistant to every therapy. The patient still had a remnant of sensibility on the sole of the foot and strong pressure of a pencil point against the sole of the foot was painful. In addition, he reported that his ulcer sometimes caused severe pain, particularly when stepping on it. At surgery, we found the tibial nerve totally transected; stimulation of the distal nerve with the strongest faradic current did not result in the slightest sensation. As I denuded the popliteal artery for the purpose of elimination of the ulcer the patient had severe piercing pains which were located in the sole of the foot. And now that the periarterial sympathetic plexus of the popliteal artery is totally extirpated, the sole of the foot is totally de-afferented, and it is entirely insensible even to the strongest pressure.

... This case therefore demonstrates that afferent nerve fibers from the sole of the foot also run through the periarterial plexus of the plantar, tibial, and popliteal arteries. However, our case tells nothing about whether the further conduction of stimuli striking the sole of the foot and experienced as pain takes place through the periarterial plexus of the popliteal, crural, and iliac arteries and of the aorta directly into the sympathetic chain; it may well take its course from the periarterial plexus of the popliteal and crural arteries via the crural nerve into the spinal cord ...

A Case of Stab Wound of the Spinal Cord, a Contribution to the Theory of the Function of the Medullary Sensory Tracts, Especially the Posterior Funiculi
Extrait du Volume Jubilaire en l'Honneur du Professeur G. Marinesco, Institut D'Arts Graphiques E. Marvan, Bucarest.

... In spite of this seemingly normal response to the painful experience, our patient offers an altogether specific and most noteworthy disturbance of pain appreciation. He is unable to distinguish whether he is pricked with a needle, strongly pulled by his hair, whether his muscles or bones are energetically squeezed, or whether a strong faradic current is applied to the skin. The same jabbing pain comes through in all instances, as long as the stimulus is strong enough. The patient cannot discriminate by any means the different varieties of pain, the different genesis of pain according to the target point of the stimulus on the skin, hair or deep receptors which any healthy individual may distinguish readily, as even our patient himself on his arms, neck and head. I have paid no attention so far to the capability of distinguishing the different types of pain according to etiology in cases of pure lesions of the posterior funiculi. In pure antero-lateral quadrant sections this capability certainly is not interfered with in any way and I would like to assume, therefore, that in our case the inability to discriminate pain is lost because of the posterior column section. Apparently in applications of peripheral painful stimuli, the posterior columns deliver to the cerebrum just those accessory excitations upon which is based the differentiation in type and genesis of pain ...

3. The Extent of the Principle of Localization

The electrical stimulation of peripheral nerves, spinal nerve roots and tracts as well as of the cerebral cortex, together with Foerster's observations on the "irritative" phenomena—that is cortically triggered seizures— and the paralyses and sensory losses following cord transections all yielded accurate knowledge of the localization of function in the nervous system. Foerster reported in detail on these matters in his Wiesbaden paper (1934). He placed particular emphasis upon the collaboration of different parts of the nervous system in the production of the motor, sensory, and vegetative effects and he indicated the importance of the loss of single "members of the cooperative". He also sketched the picture of restitution through re-organization of the remaining parts. In the Handbook of Neurology he further elaborated this most favored theme of the cooperation, disintegration and reorganization of the "cooperative", all members engaged in the neural effort, and he portrayed specifically and accurately the "synergies" in the Wernicke-Mann type of pyramidal tract paralysis.
The presentation at the Wiesbaden Congress of 1934 and the Jackson Memorial Lecture in London in 1935 were certainly the culmination of his rhetorical endeavors at meetings and of his appearance in the scientific world.

On the Significance and Comprehensiveness of the Localization Principle in the Nervous System

Verh. Dtsch. Ges. Inn. Med., XLVI Kongress 1934.

... My most esteemed colleagues: I am specially grateful to the Chairman of the German Society of Internal Medicine for the honorable assignment to discuss before this body the problem of localization in the nervous system. Wherein lies the inner connection between neurology and internal medicine? This connection is based not only historically—here I only need mention the names of ROMBERG, FRIEDREICH, ERB, SCHULTZE, KUSSMAUL, NOTHNAGEL, LICHTHEIM and STRÜMPELL—but it rests in the inner being and method of neurology. If the latter is to fulfill truly the task set before it, it will always have to keep in close touch with the mother discipline, it will have to be rejuvenated constantly by it and share in its progress ...

... If we are to assay adequately the significance and comprehensiveness of the localization principle in the nervous system, we first have to define what is to be understood by localization.

The somato-topic relationship of a certain part of the body or organ to a certain segment of the nervous system is embraced by the term localization. One may not disregard the question of the functional significance of this link. One may start by considering a certain function of the organism and by trying to determine what segments of the nervous system are taking part in the achievement of this function and what ones are indispensable for this achievement. There is no immediate identity between these two methods of consideration and a third question can be raised, i. e. how far can we take it for granted that disease of a certain system or a certain locality of the nervous system is evidenced by a certain symptom or complex of symptoms? Personally, I believe the stimulation phenomena and the loss phenomena resulting from stimulation or destruction of a certain nervous substrate are of paramount significance for the question of the functional meaning of the substrate. However, one must not consider the excitation phenomena and the deficit phenomena in isolation but rather relate them to all other pertinent phenomena. One may thereby try through synthesis to obtain insight into the physiological events ...

... Ladies and Gentlemen: I am at the end of my discourse. Naturally, I could only throw glancing lights upon the comprehensive problem of localization in the nervous system. But perhaps the lines and dots which I could show will round out for you a silhouette which permits us to recognize that the individual segments of the nervous system are not simply each other's equals, either in their somato-topic relation or in relation to certain activities of the nervous system. They may not simply be interchanged *ad libitum*, but rather there exist entirely distinct relations of some segments of the nervous system to certain performances of the organism. On the other hand, there is never one single segment of the nervous system only participating in any and every performance of the organism, but rather numerous such segments of the nervous system cooperate for every task. They are closely tied together in a cooperative, and even though through sudden elimination of one or the other member, the cooperative may initially and many times become insufficient, a readaptation to the pre-existing normal proficiency is accomplished. This occurs according to the ubiquitously recognizable principle of multiple security and by virtue of the faculty immanent in every living organism to bring into play the process of restitution after loss of substance. Often the degree of such restitution makes us bow low in amazement and admiration . . .

The Functional Analysis of the Cerebral Cortex: Motor and Sensory Cortical Fields

In his 1934 Wiesbaden paper, FOERSTER had given a synopsis of his life's work on the principle of localization. Now in the Handbook of Neurology he had another opportunity to detail his own experiences. His publisher, JULIUS SPRINGER, never imposed any restrictions as to number of pages or of pictures. Thus originated a collection of material of the first order which today must still stand as the basis of all discussions on localization in the cerebral cortex.
On the occasion of JACKSON's 100th birthday celebration, FOERSTER concisely reported on the motor cortex in a memorial lecture in London. He was particularly pleased to assume this task because he considered himself an intellectual disciple of JACKSON, whose doctrine of "levels" in the organization of the nervous system—stemming from SPENCER's philosophy—he had been trying all his life to document neurophysiologically.

The Motor Cortex in Man in the Light of HUGHLINGS JACKSON's *Doctrines*

Brain 59, 135—159 (1936).

HUGHLINGS JACKSON was the first to point out that there is such a thing as a motor cortex. He was the first to state clearly that the brain, the organ of mind, possesses motor functions. He discovered and elucidated this fact long before further evidence of the motor function of the cerebral cortex was provided by the experimental investigations of HITZIG and FERRIER on animals. JACKSON says: "The convolutions of the brain must contain nervous arrangements representing movements. There is nothing else they can represent except movements and impressions" . . .
. . . Summing up the motor disturbances resulting from destruction of the precentral convolution, we can say that the main symptoms are:

1. The *negative* symptoms are loss of the isolated innervations of single muscle groups, loss of most specialized movements, and loss of the faculty of modifying the stereotyped extrapyramidal synergies and of adjusting them to special purposive acts. The loss of function can be compensated to a greater or lesser degree by the ipsilateral precentral gyrus.

2. The *positive* symptoms are: a) Exhibition of the functions of the intact extrapyramidal cortical motor areas which are no longer under the control of the anterior central convolution and are disassociated from its activity; b) increased spinal reflex activity, the spinal reflex machinery being freed from the control exerted upon it by the anterior central convolution (Fig. 5) . . .

Symptomatology of Diseases of the Cerebrum—Motor Fields and Tracts— Cortical Sensory Fields

Handbook of Neurology, Vol. VI: General Neurology. Berlin: Julius Springer Press 1936, p. 330 ff.

. . . All motor cortex fields, the area pyramidalis (area 4) and the extrapyramidal areas, the precentral extrapyramidal area 6 a alpha, the frontal extrapyramidal area 6 a beta, the retrocentral areas 3, 1, 2, the parietal area 5 a and b and the temporal area 22 together form a working unity, all areas act in closest cooperation in the execution of movements and each area contributes its special quota to the achievement of the static and kinetic task. If even only one member should suddenly leave the cooperative, the efficiency of the total unity is interfered with initially; in this respect the different motor areas possess different dignities. In a loss of the area pyramidalis, in spite of anatomical integrity of the remaining cortical motor areas, the entire combine promptly collapses to complete incapability, voluntary motion is eliminated entirely and promptly.

. . . The area pyramidalis takes care of the modifications and supplementations of the extrapyramidal synergies that are necessary for the practical working of our limbs. It alone possesses all single components of the extrapyramidal synergies in relation to other components, through special impulses to this or that muscle group, and it may add components which *per se* are not contained at all in the extrapyramidal synergies. Second, through its inhibitory elements, the area pyramidalis is able to cancel all superfluous and purpose-hindering components inherent in the complex extrapyramidal synergies, such as the pronation component from the flexor synergy of the arm, the adduction component from the extensor synergy, the associated movements of the arm in voluntary leg motions, the associated movements of the leg in voluntary arm motions, or finally the extrapyramidal co-innervation of all muscles not involved in the task if the purpose consists in an isolated movement of an individual segment of an extremity. The pyramidal area, through its inhibitory fibers, blocks out the corticogenic extrapyramidal impulses to the anterior horn cells of all muscles not involved in the particular movement. Over and above that, by its inhibiting fibers, it prevents the extensor reflex of the antagonist from interfering with the movement in a bothersome fashion. Afferent excitations reaching the motor cortex continously from the body periphery during the process of static and kinetic efforts now play into this complicated synergy of all extrapyramidal cortical areas and the area pyramidalis, in a regulatory, stimulating or inhibiting way. Largely owing to these excitations, every muscle considered for a static or kinetic performance is actually innervated and is innervated to the required degree at any instant in the sequence of motions. On the other hand, the afferent fibers leading to the motor cortex watchfully ensure that only the muscles that are necessary for the particular task are innervated, and that no other muscles are co-innervated that would be unsuitable for or would impede the task. The afferent input is involved, just as much as the inhibitory pyramidal fibers, in the elimination and prevention of unsuitable co-movements or co-innervations . . .

. . . However, in a large number of cases of traumatic lesions of the retrocentral region, the form of the sensory defect does not follow inexorably the doctrine of the somatotopic arrangement of the postcentral gyrus as determined by stimulation experiments and confirmed by examples of sensory defects resulting from elimination of circumscribed areas of the postcentral gyrus. In the extremities particularly, the sensory defect frequently does not show the circular stocking-glove-cuff type as detailed previously, which corresponds completely to the focal arrangement in the postcentral gyrus of the individual segments of the body and the extremities. Rather, the disturbance demonstrates the form of more or less long drawn out stripes and bands, which on the one hand may involve only one side, the inside or outside, of the arm or leg, or on the other hand extend over several, even eventually all, extremity segments; fingers, hand, forearm, upper arm, toes, foot, lower leg, thigh. I have called this type of sensory disturbance the axial or longitudinal type, in contradistinction to the circular type. In the majority of cases the disturbance in the upper extremity concerns the inside, e. g., the fourth of fifth finger and the ulnar half of the hand, or the areas mentioned and the ulnar half of the forearm, maybe even the inside of the upper arm too (Figs. 48, 49 and 50). The opposite situation, that is, the outer aspect of the upper extremity remaining insensible with the inner aspect altogether spared, is much rarer . . .

... To this double arrangement in the postcentral gyrus of foci for individual body segments, on the one hand following each other in a direction from the top down, and on the other hand in anterior and posterior halves, corresponds to what I assume belongs in the arm, the radial and the ulnar half, and on the leg the lateral and medial half. This double arrangement accounts for the fact that according to the position and extent of the cortical focus the resulting sensory disturbances sometimes show the circular, sometimes the axial type. At other times, in rather mixed form, the localization of the lesions is irregular. Incidentally, I have observed repeatedly in one and the same case the axial type on the upper extremity and the circular type on the lower extremity or vica versa, the circular type on the former and the axial type on the latter.

... I am unable to indicate which subsegment corresponds to the longitudinal subdivision of the trunk area, whether the lateral and medial or the dorsal and ventral parts of the trunk. I have never been able to find the more medial sections of the trunk afflicted more severely than the lateral sections, let alone afflicted exclusively. I have no informations on this point from cortical stimulation. Sensations over the trunk are usually felt most markedly in the ventral part of the thorax and abdomen. Dissociated defects of sensation in the abdominal area are often seen in abdominal lesions. In the majority of cases the lateral parts of the trunk are more markedly or even exclusively lost. I have never observed that parts of the trunk near the midline are more markedly or exclusively affected than the lateral part. However, it occasionally happens that on the trunk the front is affected to a greater degree than the back or conversely, the latter more than the former ...

... If after this rather summary synopsis of the consequences of elimination of the cortical motor area and the available possibilities of compensation, we try to gain an idea of the cooperation of the individual cortical areas, we do best to start out from the extrapyramidal region. These set into motion entirely distinct complex movement synergies. In their character and their composition, however, these synergies represent only a poorly developed basis for the practical use of our limbs. In the flexion synergy and the extension synergy of the leg there no doubt are provided the two most important elements of locomotion. The flexion synergy of the arm doubtlessly contains in its simultaneous abduction of the upper arm and flexion of the forearm two important constituents of all those composite movements of the arm by which the hand is brought to the mouth, to the head or to the back. In the extension synergy of the upper extremity, the extension of the forearm, there is contained at least one part of a motion important for the grasping of an object found before us ...

The development of these ideas on synergy patterns and their comparison with phylogenetically archaic patterns of movement had been presented by FOERSTER in earlier papers.

The Phylogenetic Element in Spastic Paralysis

Berl. klin. Wschr. *26* and *27* (1913).

... The specific subcortical posture and the specific subcortical movements of the limbs, as I have termed them, and which we find in the newborn child and in spastic paralyses have a phylogenetic significance ...

... By comparison we see that the spastic contractures imitate the climbing posture of the monkey at rest, limb by limb. The limb posture shows in essence the same components in both. As far as the big toe is concerned, the abduction position in the monkey corresponds to the extension of the first phalanx in the human. At times also the big toe stands, in fact, abducted in congenital diplegias ...

... Likewise, the motion synergies in spastic paralyses have a great and decided similarity to climbing movements of the monkey.

... Certain reminiscences of the climbing are found almost always in such cases of spastic paralyses in which upright gait is essentially preserved. This holds, as mentioned above, particularly for supination of the foot, for the claw position of the toes, the inward rotation and adduction of the leg in forward extension, for the associated movements of the arms and for several other details.

4. Contributions to Neurophysiology

a) The First Electrocorticogram of a Brain Tumor

Electrical stimulation of the cortex as it had been carried out in the "classical" neurophysiology of FRITSCH and HITZIG, of FERRIER, FEDOR KRAUSE, Sir VICTOR HORSLEY and SHERRINGTON was also for FOERSTER one of the principal methods for the analysis of function of the nervous system. FOERSTER employed immediately the more modern methods of picking up action currents from the brain, which had just been published by BERGER. He and his co-worker, ALTENBURGER, were the first clinicians to employ this method on the exposed human brain. Thus, FOERSTER was the first to have, among other innovations, outlined a brain tumor corticographically during surgery. He saw amplitude reduction and demonstrated slowing of wave activity. In 1935 he was able to describe the results of 30 corticographies elicited under various circumstances. He found no potentials engendered by peripheral stimulation (evoked potentials), but he did see the electrophysiological correlates of a focal seizure. Also today's routinely employed forced hyperventilation during the taking of an EEG tracing goes back to FOERSTER and was suggested by him for the provocation of epileptic seizures.

Electrobiological Processes in the Human Cortex

with H. Altenburger. Proceedings of the Association of German Neurologists, 22nd Anniversary, Munich, 1934. Berlin: Vogel Press 1935, p. 93–102.

... If we wish not only to demonstrate the fundamental bioelectric behavior of the cortex, but also desire to gain insight into the mode of function of certain areas, the leads have to be placed upon the cerebral cortex itself. Furthermore, unipolar leads are necessary; that means we have to place one electrode at a site which is practically free from oscillations in potential—for this we chose the ear lobe—and the other on that portion of cortex proper that is to be examined. We wish to report briefly the results of such observations in 30 patients

... At the point recorded in the picture, a cutaneous stimulus was active, a stroke over the sole of the foot; no change in the wave pattern is manifested. The amplitude and the frequency of potentials remain the same ...

... Even in active motion such as fist closure, we have so far been unable to establish changes in the wave pattern in the vicinity of the frontal area 6 a beta, and the same holds for psychic efforts as shown in the following picture (Fig. 4) ...

... Tracings were made of the leg area of the precentral gyrus in the course of an operation for epilepsy (Fig. 7) during a series of clonic discharges. One sees that these are accompanied by a rapid sequence of potential changes of great amplitude ...

... In conclusion, several observations from pathology ought to be mentioned. We have noticed on a number of occasions in brain tumors that in the tumor area or its vicinity there were only very small potentials. Thus in the tracing shown (Fig. 11), which is one of a frontal brain tumor, the potentials are conspicuously low, which is very clearly seen when compared with pictures of the normal frontal lobe activity. The subsequent tracing (Fig. 12) of a parietal tumor shows an entirely analogous picture. The potentials are conspicuously low ...

b) Hyperventilation as a Method to Provoke Seizures

Hyperventilation Epilepsy

Zschr. Nervenhk. 83, 347–356 (1924).

... And on the basis of the fact one may succeed in eliciting artificially the typical tetany syndrome through forced breathing in the healthy, I have carried out systematic examinations of the influence of hyperventilation in 45 patients with epileptic seizures.

These experiments were carried out in identical fashion and the patients, in a sitting position, were induced to breathe intensely for 10 minutes. The main emphasis is to be put upon forced respiration, and careful watch has to be kept for the occurrence of spontaneous motor excitatory manifestations. In short, the results of the examinations are as follows: 1) of 45 epileptics, an epileptic attack occurred during hyperventilation in 25, that is in 55.5% of the cases . . .

c) Reflex Physiology

As soon as FOERSTER had enlisted H. ALTENBURGER as a co-worker, he was able to have checked electrophysiologically his clinically demonstrated opinions on movement, reflexes, spasm, hypotonia, rigor, etc. A series of papers show the layout of this work, the modus operandi of which should at least be touched upon here.

Contributions to the Physiology and Pathophysiology of the Tendon and Bone Phenomena in the Pyramidal Tract Syndrome

With H. ALTENBURGER. Zschr. ges. Neurol. Psychiat. *147*, 779–790, (1933).

. . . One of us (FOERSTER) some time ago produced what, in our opinion, is the only certain proof for the existence of a true reflex spread. In a spastic cortical leg paralysis, where numerous muscles of the paralyzed contralateral leg (quadriceps, adductors *et al)* could be brought to contraction by tapping the tendons and bony protuberances of the healthy leg, the posterior roots, L2 to S1, of the paralyzed extremity were transected. Afterwards, no quadriceps or adductor reflex could be elicited by tapping the paralyzed extremity, but jerking of these muscles of the extended leg occurred promptly when the patellar tendon or the medial epicondyle of the unextended, healthy side was tapped. No transmission by mechanical percussion could be considered because the muscles concerned were totally bereft of their sensory supply. The participation of the contralateral muscles in the reflex achievement could only be based upon a spread of the reflex through nervous channels . . .

On the Physiology and Pathophysiology of the Tendon and Bone Phenomena in Relation to the Stretch Reflexes

With ALTENBURGER. Zschr. ges. Neurol. u. Psychiat. *156*, 478–483 (1936).

. . . In typical cases, hypotonia is electrophysiologically characterized by the fact that upon passive stretching, even if effected as suddenly as possible, the muscles involved remain without current. At the same time, the adaptation reflex fails in those muscles whose insertion points are approximated (Fig. 3).

d) FOERSTER'S *Opinions on the Activating and Inhibiting Systems of the Brain Stem*

The activating influence of the "brain stem" upon the "psyche" especially interested FOERSTER. He reports in a dramatic picture of a surgical experience the triggering of a "maniacal" syndrome through stimulation of anterior portions of the brain. It is too little appreciated that OTFRID FOERSTER had very accurate ideas concerning the localization of the activating and inhibiting systems. As early as 1934 he spoke to the matter in his great report on the "Comprehensivenes of the Localization Principle". He elaborated upon it later in a paper with GAGEL and MAHONEY.

On the Significance and Comprehensiveness of the Localization Principle in the Nervous System

Verhandl. Dtsch. Ges. Innere Medizin, XLVI Kongress, Wiesbaden 1934.

. . . The second important fact for our problem is that through mechanical influence upon the anterior portions of the hypothalamus, such as occurs in operative interventions on tumors of the chiasmatic region compressing the infundibulum from below and which we always approach surgically by way of a transfrontal route, there develops the opposite psychic condition, that is a great agitation, a typical maniacal syndrome with an expansive urge to move and talk, accompanied by an outspoken flight of ideas. In one of my patients with a suprasellar craniopharyngioma pushing against the floor of the third ventricle from below, when I arrived at the tumor in the above-mentioned way, the most violent mania started at the instant I began to manipulate the tumor. The patient went through a veritable logorrhea, quotations from the Latin, Greek and Hebrew languages tumbled forth in a wild melee. He reacted upon every word I dropped with a flight of ideas: as I asked for a "Tupfer", he reacted: Tupfer, Tuepfer, Hupfer, hüpfen Sie mal. As I demanded the "Messer" he reacted: "Messer, Messer, Metzger, Sie sind ein Metzger, ein Metzel, das ist ja ein Gemetzel, metzeln Sie doch nicht so, messen Sie doch, Sie messen ja nicht, Herr Professor, profiteor, professus sum, profiteri." All the maniacal reactions were in strict dependance upon manipulation of the tumor proper and its influence upon the floor of the third ventricle. They were not elicited from any other location in the neighborhood.

The encephalic Tumors of the Medulla Oblongata, the Pons and the Midbrain
With O. GAGEL and W. MAHONEY. Arch. Psychiat. Nervenkr. *110*, 1–74 (1939).

... From the anterior portion of the hypothalamus the cortex is turned on like an electric light ... the stimulation of this trigger area produces increased psychic activity ... paralysis, that is turning off ... produces a desire to sleep, stupor, unconsciousness. Contrariwise, from the caudal segment of the brain stem, the oblongata, the pons, the midbrain, the light ... in the cortex being switched off, cortical activity is inhibited and paralyzed. Stimulation of these caudal brain segments ... produces fatigue, a desire to sleep, stupor, unconsciousness, coma ... The alternating change from stimulation to inhibition manifests itself in changing conditions of maniacal agitation on the one hand and somnolence, sopor, unconsciousness on the other.

A Case of So-called Glioma of the Optic Nerve—Spongioblastoma Multiforme Ganglioides
With O. GAGEL. Zschr. ges. Neurol. Psychiat. *136*, 335–336 (1931).

... Initially, our patient showed alternating states of maniacal excitation and torpid passivity. We have observed this same change in other cases of tumor in the floor of the third ventricle. Later on a prominent feature was a practical total turn-about of the diurnal schedule, in which the patient slept mostly during the day, and during the night regularly presented a severe agitated Korsakow-like picture, which was also noted promptly during the day when the patient was awakened.

e) *Contributions to the Knowledge of the Vegetative System*

His work on tumors of the pituitary and hypothalamus provided FOERSTER with an accurate knowledge of the symptomatology of the diencephalon, which he delighted to protray in glowing colors and rather three-dimensionally, albeit sometimes rather dramatically. FOERSTER occupied himself extensively with the anatomy of the vegetative system, stimulated by his wide operative experience. I have mentioned that he probably was the first to perform a supradiaphragmatic resection of the splanchnic nerves–judging from a letter of FOERSTER to KAHN–(ADSON'S operation). His synopsis of the "radicular sensory supply of the viscera" is still a landmark (Handbook of Neurology, Vol. 3, p. 290, 1936). Other studies of his on vegetative regulation are also interesting, especially that on thermoregulation, where the vasomotor tract in the lateral funiculus (LOEWENTHAL'S bundle) is described (p. 56). Finally, he and his colloborators worked on pupillary innervation in an experimental and morphological study, providing a typical example of the style of work of that time in Breslau: Clinical observations and question asking, experimental control in the monkey, with test of functions after appropriate nerve section, morphological investigation in the human and in the experimental animal, final suggestions for diagnosis and therapy.

Vegetative Regulation

With O. GAGEL and W. MAHONEY. Verh. Dtsch. Ges. Inn. Med., pp. 165–187 (1937).

... The patient in question was already receiving triple rations, and in addition he avidly devoured all the leftovers he could procure from other patients. There were severe visual difficulties, an advanced grade of diabetes insipidus and a pronounced mania, manifesting itself in continuous euphoria and unbridled logorrhea. The many smart and jovial remarks which the patient emitted in his thin and bird-like child's voice, appeared sometimes even comical. This symptom complex was conditioned by a large craniopharyngioma ...

... The second patient, on completing the 7th year of his life, suddenly started to grow rapidly and to put on weight so that he weighed 42 pounds at $7^1/_2$ years. That means he had gained almost as much weight in the last half year as he had in the seven preceding years. Simultaneously with this sudden acceleration of growth there developed severe visual disturbances, increasing polydipsia and polyuria, and sexual precocity. This $7^1/_2$ years old boy (Fig. 20, p. 180) was one of those precocious little bucks of whom I have spoken previously, who repeatedly, energetically and indiscriminately attacked adult persons of the female sex of various ages and various appearances ...

... The voice was a full sounding male voice, if one disregards an occasional wavering into childish ranges ...

... Contrary to the monkey, ligation of the pituitary stalk is followed regularly by diabetes insipidus. What is the basis for this difference between the monkey and the dog? So very often in experimental physiology one emphasizes with justification: no premature conclusions from one animal to the other, particularly none such from animal to man. However, in our case the secret lies in a different direction: anatomy first and then physiology, and if physiology first, then not without anatomy. Fig. 3 shows you the pituitary stalk in the monkey; it is long. It offers itself for ligation without concomitant damage to neighborhood structures ...

On Disturbances of Thermoregulation in Diseases of the Brain and Spinal Cord and in Surgery of the Central Nervous System

Jb. Psychiat. Neurol. Vol. 52, Fascicle 1, 1–14 (1935).

... For the most part, the thermoregulatory tracts descending from the brain stem to the spinal cord run in the anterolateral funiculus. In a transection of the anterolateral funiculus at the cervical level there may occur hyperthermia, similar to the hyperthermia seen in total transection of the medulla. As a rule, however, the vasodilatation is not nearly so pronounced as in a total transection of the medulla. The degree of hyperthermia depends on the extent of the transection of the vasoconstrictor tract, which lies tucked away in the corner between anterior and posterior horn and immediately in front of the pyramidal tract . . .

On the Anatomy, Physiology and Pathology of the Pupillary Innervation

With O. GAGEL and W. MAHONEY. Verh. Dtsch. Ges. Inn. Med. XLVIII. Kongr., p. 394 (1936).

. . . The superior cervical ganglion receives its stimuli from the spinal cord, that is from the ciliospinal center lying in the 8th cervical and the 1st and 2nd thoracic segments, which center belongs to the sympathetic lateral horn (Fig. 7). The pupillary fibers pass through the corresponding anterior roots, C8, T1, and T2. The preganglionic myelinated fibers run through the corresponding white rami communicantes into the cervical chain and up in it as far as the superior cervical ganglion, where they synapse with the cells of the latter. After extirpation of the superior cervical ganglion, there occurs retrograde tigrolysis of the lateral horn cells within the segments C8, T1, and T2, but in no lower segment (Figs. 8, 9 a and b).

. . . Electrical stimulation of the 8th anterior cervical and the 1st and 2nd anterior thoracic roots in the human causes pronounced pupillary dilatation and exophthalmos. The situation in the human is such that the stimulation effect from T1 is very constant and always the strongest. If in addition there is a good response from stimulation of C8, then as a rule response from T2 is absent, and conversely the effect from stimulation of C8 is missing if T3 causes a strong dilatation. Therefore, in the realm of spinal pupillary innervation there are also two types, the prefixed, more oral and the postfixed, more caudal types . . .

5. FOERSTER'S Interest in the Physiology of Special Senses

The modern physiology of the special senses, which became the main field of work of the Breslau Neurological Institute under his successor, V. v. WEIZSÄCKER, held very little interest for FOERSTER. If any problem in this area intrigued him, it was only because of its localizing value. Thus he occupied himself with the anatomy of the visual pathways, and specifically with the cortical representation of the macular region.

It was only on one other occasion when, stimulated by his co-worker, LOEWI, he attacked a problem dealing with the physiology of the special senses, and he was able to demonstrate a very interesting result. He proved that knowledge of the physiology of the special senses facilitates the physiological analysis of a complicated stimulus. However, FOERSTER did not envision a higher level of physiology for the special senses, let alone a level of Gestalt psychology, but rather looked on their function as a pure electro-physiological process, the effect of which emanated from the sensory sphere onto the sphere of perception.

On the relation of Idea and Impression in Cases of Injury to Afferent Conductive Tracts

With M. LOEWI. Zschr. ges. Neurol. Psychiat. 139, 658–693 (1932).

... In our investigations we have shown the facilitory influence of the cortical stimulation which is associated with a particular image—an influence which, acting on a cortical primary sensory area, lowers the perceptive level. In the report that we have given it may be noted that there is preparation of a special perception by the formation of an image corresponding to the reception of this cortical area ...

... No doubt a fixed cortical stimulation complex corresponds to the image of each individual part of the body. Obviously through this stimulation complex, which leads to the image of an extremity, the cortical sensory area is being conditioned again, as in the examples of our first series of experiments. The difference lies in the *in toto* conditioning of a certain segment of the cortical sensory area, that is of the one corresponding to that part of the body where the receptors are stimulated, and not in the conditioning for afferent excitation of a certain quality as in the first series of experiments. Excitations flowing promptly from the receptors to this segment of the sensory field find a prepared situation ...

... An image exercises no "effect" upon a perception and vice versa. Whoever imagines psychic phenomena as centers of activity able to act upon one another, neglects the most fundamental principle of psychic processes. And naturally, this fundamental principle of all psychic processes dominates imagination and perception. However, one cannot say: The imagination influences the perception. Here what one erroneously terms "influence" of the imagination upon the perception, we may express precisely, methodically and correctly by saying: I influence my perception by my imagination. The so-called influence of the imagination upon the perception is a psychological problem only where it emanates from the ego; in short, where it signifies as much as "I influence". And likewise, "correlation between imagination and perception" means "I relate my imagination to my perception".

Investigations of the Visual Pathways

Stimulated also by his ophthalmological colleague, LENZ, in Breslau, FOERSTER frequently occupied himself with the course of the visual pathways, and reported on it in 1929 on the basis of four cases of accurate examinations. He checked his thesis over and over again against the clinical pictures of his patients with brain tumors and he finally summarized them in his 1934 report on localization. The macular fiber pathways interested him particularly.

A Case of Recklingshausen's Disease with Five Different Coexisting Tumor Formations

With O. GAGEL, Zschr. ges. Neurol. u. Psychiat. *138*, 339–360 (1932).

... If one looks at Figure 4, representing a horizontal section through the striate area, and one sees to what a considerable extent it is destroyed, it appears miraculous that no clinically demonstrable visual defect corresponded to this destruction. However, on closer examination this discrepancy becomes understandable. The destroyed portion of the striate area corresponds to its posterior part, in which part of the calcarine area we have the representation of the macula. The macula is represented in both hemispheres, according to the studies of WILBRAND and LENZ and others, to whom we have referred in previous papers. The elimination of the posterior segment of the left striate area is fully compensated for by the right striate area. The findings in our case represent in our opinion a new and important proof of the doctrine of bilateral representation of the macula ...

IV. Congress Speeches and Biographical Appreciations

The great addresses that FOERSTER gave at the opening sessions of the Society of German Neurologists are well known. He often took weeks to inform himself on the history and academic tradition of the university or town at which the Society was meeting, recounting them in detail in his speeches. He knew the biographies of all the scientists of the city of the meeting, he sketched the cultural and scientific significance of the meeting place. Embellished by quotations from numerous poets, he created small masterpieces of classical rhetoric. In one such congress speech, he dealt with the position of neurology in internal medicine and he emphasized the necessity of training in internal medicine (1928).
The extent of preparation for such congresses is shown in his speech (1927) in Vienna. There he reported in detail about TÜRCK, NOTHNAGEL, MEYNERT, ANTON, KRAFFT-EBING, OBERSTEINER, WAGNER V. JAUREGG, MARBURG, BENEDIKT, BRÜCKE, ROKITANSKY and representatives of all the clinical schools and their less well known collaborators. More than 100 names were mentioned whose bearers somehow, at least in a sentence, were accorded recognition of their scientific significance.

From a letter of Nonne to Mrs. ILSE FOERSTER-ROSENFELD

... For eight years your father gave the first speech as his official duty as chairman. These were always masterpieces in which the earlier and current neurologists of the host city were recognized. In addition, light was shed upon the history of the city or province and its contributions to culture and science. These speeches, which were initially given by HERMANN OPPENHEIM and later on by me for eight years when we were chairmen, were presented by OTFRID FOERSTER in an accomplished form, intensely captivating in their content, with many witty quotations. Everyone listened attentively. As with HOCHE at the annual meeting in Baden-Baden, one felt how he sensed himself to be in his element, and how much pleasure it gave him to tell us about the respective places ...

Opening Address on the Occasion of the 18th Annual Meeting of the Society of German Neurologists in Hamburg, 1928

Verhdlg. d. Ges. Dtsch. Nervenärzte, 18. Jahresvers., P. 4–28. Leipzig: F. C. W. Vogel Press 1929.

... However, there is a horizon of responsibility for neurology if it desires to attain independence and maintain it. The neurologist not only has to master the anatomy, physiology and pathology of the nervous system, but he also should be his own thorough technican in all diagnostic methods ...
... On the other hand, neurology has the obligation to maintain the closest touch with basic medicine and with all its special disciplines, if it is to interpret its position correctly within medicine as a whole, and if we are to remove that justification for the reproach levelled against specialization, that it forgets about the diseased organism in its concern for the sick organ ...

. . . I should like to limit myself to a few remarks concerning the contact which neurology must have and keep with internal medicine. The neurologist must know how far a disease picture that he encounters—how far symptoms that are complained of—are an expression of a general disease of the entire organism, or can be a consequence of a disease of this or that internal organ. How often is a so-called arm neuralgia merely the expression of myocardial disease, now often aortitis, how often is a headache merely a symptom of chronic renal disease or hypertension, how often is combined system disease an indication of the beginning of pernicious anemia, how often are sciatica or a nuisance pruritus due to latent diabetes? How often will the neurologist, being honest in counselling his patient conscientiously, have to say: Yout don't belong here with me but rather with the internist?

. . . However, how often will the neurologist have to appeal for the help of the internist when it comes to analyzing a certain deviation in the function of this or that internal organ or of certain metabolic disorders in primary neurological disease. Here I think, for instance, of the disturbances of the secretory and motor functions of the stomach and the very peculiar disturbances of water balance and the total metabolism in gastric crises. I think of the disturbances of blood sugar content, water balance, and salt tolerance in diseases of the pituitary or in tumors of the region of the tuber cinereum, especially in space occupying lesions associated with increased intracranial pressure. I think of the disturbances of thermo-regulation in operative interventions on the cerebrospinal fluid system or in the area of the upper cervical cord. I beg the gentlemen from internal medicine not to regard it as a matter of chance that we have chosen the vegetative nervous system as the central focus of this year's meeting. We mean to show that our specialty wants to stand on the grounds of general medicine. We neurologists want to learn from anatomy, physiology and pharmacology as well as from internal medicine, just as we want to present what we on our part have to say on the influence of the nervous system upon the function of individual organs of the body. The closest cooperation of our speciality with all other branches of medicine shall be our motto . . .

HARVEY CUSHING

Zbl. Neurochir. 4, 195–197 (1939).

. . . CUSHING'S instinctive reservation toward brain tumors of delicate location (quadrigeminal plate, third ventricle, aqueduct) is explained from this holy respect. Not that his surgeon's masterhand would basically have recoiled. On the contrary, he proved in numerous cases that even they can be dealt with through a bold and certain attack. But he realized with his clear, critical vision that a "successful" removal of the lesion in the vast majority of deep seated brain tumors fails because of the site of the tumor and its bleeding into surrounding vital brain structures ...

... This gigantic effort of CUSHING's has borne the richest fruit. Today, a meningioma of the olfactory groove, an eosinophilic adenoma of the pituitary, an acoustic neuroma, an astrocytoma of the cerebellum or a medulloblastoma of the cerebellum may be diagnosed with almost automatic certainty by the respective symptomatology and the age of the patient. Prior to CUSHING, there was no discussion, nobody even thought of such a possibility. Tumor classification, executed in the minutest detail, is the life work of HARVEY CUSHING, not only from a pathological point of view but above all from the standpoint of practical therapeutics.

OTFRID FOERSTER *on the Occasion of the Foundation of the First Neurosurgical Specialty Journal, the "Zentralblatt für Neurochirurgie", Ed.* W. TÖNNIS

... Granted, neurology maintains the most intimate relation to all branches of medicine as almost no other specialty does and the goals of neurosurgery do not lie in the surgical treatment of nervous disease alone. In addition, it includes all those surgical interventions on the nervous system that provide treatment for the most variegated diseases of other organs. One may mention anterior quadrant section for the elimination of painful conditions of various origins, interventions upon the sympathetic system and upon the roots of the spinal cord, and the funiculi of the spinal cord in obliterating vascular diseases, in angina pectoris, hypertension, in numerous ulcer formations resistant to any other therapy, in scleroderma and in many other diseases. But even though the distribution of neurosurgical papers over the various neurological, surgical, medical, pediatric, ophthalmological and otological journals reflects the multiplicity and the manifold mutual relations of neurosurgery to all other disciplines, one must agree, on the other hand, that a condensation in a special neurosurgical publication can only advance the further development of neurosurgery proper. Such a condensation is clearly mandatory ...

Recommendations of the Bern Congress (1931) to the Governments of Countries

Following the reports and the discussion on "The relation of neurology to medicine in general and to psychiatry at the universities and hospitals of the various countries", the following resolution, suggested by Professor OTFRID FOERSTER, was passed unanimously.

Proceedings of the First International Neurological Congress Bern 1931. Bern: Verlag Staempfli & Co. 1932.

Resolution

From the Proceedings of the First International Neurological Congress Berne, August 31 to September 4, 1931

"Neurology represents an entirely independent speciality in Medicine. Unfortunately, this fact has not been sufficiently recognized in various countries. The First International Neurological Congress hopes that the universities and hospital authorities of the various states will take active steps to further the progress of Neurology."

Epilogue

Foerster, like Nonne, always argued for the special position of neurology as a separate academic branch of medicine. He proclaimed this special position before the entire world, drafting for the First International Neurology Congress in Bern in 1931 a resolution addressed to the governments of the world. Since then many countries, especially the United States and in recent years both parts of Germany, have recognized this appeal. Foerster would have been happy had he lived to see the Foundation of so many chairs of neurology in Germany's Universities and Medical Academies. His efforts for neurology have played a decisive part in the past 25 years in the development of international neurology. His work is not forgotten and one may rightly say, that his contributions continue to influence the discipline, though it is especially regrettable that his methodology has found so little following and appreciation in German neurology, neurosurgery, and neurophysiology. We find scarcely any papers in the German Neurological Archives on the doctrine of localization. In similar fashion one sees in the English speaking literature the diminishing impact of Fulton's work. May this contact with Foerster's contribution redirect German efforts toward a consideration of functional localization.
The German Society for Neurology and the German Society for Internal Medicine remember well this man who fructified world science far beyond the boundaries of his own country. May his work be an example for the young who wish to dedicate themselves to neurology.

OTFRID FOERSTER Received the Following Honors:

1. Erb Medal 1920
2. Möbius Medal 1920
3. Nothnagel Memorial Medal 1932
4. Cothenius Medal 1935
5. Hughlings Jackson Medal 1935

In addition he was honored by memberships in the following scientific societies

1. Honorary member and honorary chairman of the Gesellschaft Deutscher Nervenärzte
2. Curator of the Kaiser-Wilhelm Institut für Hirnforschung, Berlin

3. Senator of Kaiserl. Leopoldinische Akademie der Naturforscher in Halle
4. Member of the scientific committee of the Gesellschaft Deutscher Naturforscher and Ärzte
5. Honorary member, Wiener Verein für Psychiatrie und Neurologie
6. Honorary member, Wiener Gesellschaft für Innere Medizin
7. Corresponding member, Wiener Gesellschaft der Ärzte
8. Honorary member, Deutsche Ärztevereinigung Prag
9. Honorary member, Royal Society of Medicine, London
10. Corresponding member, British Medical Association
11. Honorary member, British Association of Neurological Surgeons
12. Corresponding member, Societé de Neurologie de Paris
13. Honorary member, American College of Surgeons 1930
14. Honorary member, American Neurological Association
15. Honorary member, Academy of Medicine, New York
16. Corresponding member, Deutsche Ärztegesellschaft, New York
17. Surgeon in Chief pro tempore Peter Bent Brigham Hospital Boston 1930
18. Corresponding member, Harvey Cushing Society
19. Honorary member, Academia Medica di Roma
20. Corresponding member, Italienische Gesellschaft für Neurochirurgie
21. Honorary member, Türkische Ärztegesellschaft
22. Honorary member, Jugoslawische Oto-Neuro-Ophthalmologische Gesellschaft
23. Honorary member, Moskauer Pathologische Gesellschaft
24. Honorary member, Ukrainische Ärztegesellschaft
25. Honorary member, Estnische Gesellschaft für Neurologie
26. Corresponding member, Warschauer Neurologische Gesellschaft
27. Corresponding member, Societas Medicorum Sudeana
28. Corresponding member, Königlich Schwedische Gesellschaft der Naturforscher Lund
29. Honorary member, Kopenhagener Ärztegesellschaft
30. Honorary member, Holländische Gesellschaft für Psychiatrie und Neurologie

Bibliography

1897

Quantitative Untersuchungen ueber die agglutinirende und baktericide Wirkung des Blutserums von Typhus-Kranken und Reconvalescenten. Inaugural-Dissertation, Med. Facultät, Universität Breslau, 20. Mai, 1897.

Quantitative Untersuchungen über die agglutinirende und bactericide Wirkung des Blutserums von Typhuskranken und -Reconvalescenten. Z. Hyg. InfektKr. *24*, 500–529 (1897).

Die Serodiagnostik des Abdominaltyphus. Fortschr. Med. *15*, 401–409 (1897).

1899

Les troubles de la sensibilité dans le tabes. Rev. neurol. *7*, 822–826 (1899) (mit H. S. Frenkel [1]).

1900

Untersuchungen über die Störungen der Sensibilität bei der Tabes dorsalis. Arch. Psychiat. Nervenkr. *33*, 108–158, 450–520 (1900) (mit H. S. Frenkel [1]).

Zur Symptomatologie der Tabes dorsalis im praeataktischen Stadium und über den Einfluss der Opticusatrophie auf den Gang der Krankheit. Mschr. Psychiat. Neurol. *8*, 1–14, 133–150 (1900).

Demonstration: Fall von einem eigenthümlichen psychischen Zwangsphänomen, etc. (78. Sitz. Ver. ostdtsch. Irrenärzte, Breslau. 24. Feb. 1900) Allg. Z. Psychiat. *57*, 411–414, 415 (1900). Ref. Rev. neurol. *9*, 258 (1901).

1901

Uebungstherapie bei Tabes dorsalis. Dtsch. Ärzteztg., 100–104, 128–131 (1901).

Untersuchungen über das Localisationsvermögen bei Sensibilitätsstörungen. Ein Beitrag zur Psychophysiologie der Raumvorstellung. Mschr. Psychiat. Neurol. *9*, 31–42, 131–144 (1901).

Beiträge zur Physiologie und Pathologie der Coordination. Die Synergie der Agonisten. Mschr. Psychiat. Neurol. *10*, 334–347 (1901).

1902

Die Physiologie und Pathologie der Coordination; eine Analyse der Bewegungsstörungen bei den Erkrankungen des Centralnervensystems und ihre rationelle Therapie. Jena, G. Fischer, 1902. xiv, 318 pp. 8°.

Ein Fall von Poliomyelitis im obersten Halsmark. (Med. Sekt. schles. Ges. vaterl. Kult. Breslau. 6. Dez. 1901.) Allg. med. ZentZtg. *71*, 13–14 (1902).

Discussion, Mams: Ueber die Frühdiagnose der Tabes mit besonderer Berücksichtigung der Augensymptome (Med. Sect. schles. Ges. vaterl. Kult. Breslau. 4. Juli 1902.) Allg. med. ZentZtg. *71*, 714 (1902).

Ueber einige seltenere Formen von Krisen bei der Tabes dorsalis sowie über die tabischen Krisen im Allgemeinen. Mschr. Psychiat. Neurol. *11*, 259–283 (1902).

1903

Atlas des Gehirns. Schnitte durch das menschliche Gehirn in photographischen Originalen. Abt. 3. –

21 Sagittalschnitte durch eine Grosshirnhemisphäre. Herausgegeben von Carl Wernicke. Breslau, Verlag der Psychiatrischen Klinik (1903).

Die Mitbewegungen bei Gesunden, Nerven- und Geisteskranken. Jena, G. Fischer (1903). 53 pp. 8°. Ref. Rev. Neurol. *12*, 555 (1904).

Beiträge zur Kenntnis der Mitbewegungen. Jena, G. Fischer, 32 pp. (1903).

Ein Fall von elementarer allgemeiner Somatopsychose (Afunktion der Somatopsyche). Ein Beitrag zur Frage der Bedeutung der Somatopsyche für das Wahrnehmungsvermögen. Mschr. Psychiat. Neurol. *14*, 189–205 (1903).

Ueber Uebungstherapie (Med. Sect. schles. Ges. vaterl. Kult. Breslau. 5. Dec. 1902). Allg. med. ZentZtg. *72*, 15 (1903).

Vergleichende Betrachtungen über Motilitätspsychosen und über Erkrankungen des Projektionssystems. Antrittsvorlesung. Habilitation als Privatdozent f. Nervenheilkunde und Psychiatrie, 10. Aug. 1903.

1904

Die Fasersysteme des Grosshirns des Menschen. Arch. Psychiat. Nervenkr. *39*, 924–928 (1904).

Ein Fall von Dementia paralytica nach Typhus abdominalis mit Ausgang in vollkommene Heilung. Mschr. Psychiat. Neurol. *16*, 583–589 (1904). Ref. Neurol Zbl. *24*, 81 (1905).

Grundlagen der Übungsbehandlung bei der Hemiplegie (76. Versamml. Ges. dtsch. Naturforsch. Ärz. 1904). Verh. Ges. dtsch. Naturf. Ärz., 308–310 (1904).

Hirnveränderungen bei Erschütterung (76. Versamml. Ges. Naturforsch. Ärz. 1904). Verl. Ges. dtsch. Naturf. Ärz., 525–528 (1904).

Das Wesen der choreatischen Bewegungsstörungen. Samml. klin. Vortr., 1904, No. 382 (Inn. Med. No. 113), 259–294. Ref. Neurol. Zbl. *24*, 912 (1905).

Compensatorische Übungstherapie bei der Tabes dorsalis. Lehrbuch der physikalischen Heilmethoden. Wien, 1904.

1905

Zwei Fälle von Friedreich'scher Krankheit (Med. Sect. schles. Ges. vaterl. Cult. Breslau. 3. März 1905). Allg. med. ZentZtg. *74*, 232 (1905).

1906

Die Kontrakturen bei den Erkrankungen der Pyramidenbahn. Berlin, S. Karger, 1906. 65 pp. 8°. Ref. Rev. Neurol. *14*, 1157 (1906).

Demonstration eines Falles von hysterischer Bewegungsstörung im Bereiche des linken Augenlides. Allg. Z. Psychiat. *63*, 339–343 (1906).

Ein Fall von isolierter Durchtrennung der Sehne des langen Fingerstreckers. Ein Beitrag zur Physiologie der Fingerbewegungen. Beitr. klin. Chir. *50*, 676–683 (1906).

1907

Erfahrungen über die Behandlung von Störungen des Nervensystems auf syphilitischer Grundlage. Neisser Festschrift. Arch. Derm. Syph., Wien, *86*, 3–44 (1907) (mit Harttung [1]).

Ein Fall von Cysticerkus der Gehirnrinde durch Operation entfernt (Med. Sect. schles. Ges. vaterl. Kult. Breslau. 15. Nov. 1907). Allg. med. ZentZtg. *76*, 782–783 (1907).

Fall von Commotio und Contusio cerebri: Aphasie, Rindenepilepsie, Trepanation, Heilung (Med. Sect. schles. Ges. vaterl. Cult. Breslau. 15. Nov. 1907). Allg. med. ZentZtg. *76*, 790 (1907).

Demonstration: a. Cysticercus im Gehirn. b. Epilepsie nach Trauma (Med. Sekt. schles. Ges. vaterl. Kult. Breslau. 15. Nov. 1907). Dtsch. med. Wschr. *33*, 2199 (1907).

Diskussion, ANSCHÜTZ: Beitrag zur Chirurgie des Kleinhirntumors (Med. Sekt. schles. Ges. vaterl. Kult. Breslau. 23. Nov. 1906). Allg. med. ZentZtg. *76*, 10–12 (1907).

1908

Drei Fälle von isolierten Sehnenverletzungen. Ein weiterer Beitrag zur Physiologie und Pathologie der Fingerbewegungen. Beitr. klin. Chir. *57*, 720–733 (1908).

Ueber eine neue operative Methode der Behandlung spastischer Lähmungen mittels Resektion hinterer Rückenmarkswurzeln. Z. orthop. Chir. *22*, 203–223 (1908).

Demonstration: Ein Fall von linksseitiger Kleinhirncyste, operativ entfernt. Heilung (Med. Sekt. schles. Ges. vaterl. Kult. Breslau. 31. Jan. 1908). Allg. med. ZentZtg. *77*, 130 (1908).

1909

Beiträge zur Hirnchirurgie. Berl. klin. Wschr. *46*, 431–436 (1909).

Zur Symptomatologie der Poliomyelitis anterior acuta. Beobachtungen während der diesjährigen Epidemie in Breslau. Berl. klin. Wschr. *46*, 2180–2184 (1909).

Über den Lähmungstypus bei cortikalen Hirnherden. Dtsch. Z. Nervenheilk. *37*, 349–414 (1909).

Ueber operative Behandlung gastrischer Krisen durch Resektion der 7.–10. hinteren Dorsalwurzeln. Beitr. klin. Chir. *63*, 245–256 (1909) (mit H. KÜTTNER). Ref. Berl. klin. Wschr. *46*, 2031 (1909).

Die arteriosklerotische Muskelstarre. Allg. Z. Psychiat. *66*, 902–914 (1909).

Ueber die Behandlung spastischer Lähmungen mittels Resektion hinterer Rückenmarkswurzeln. Mitt. Grenzgeb. Med. Chir. *20*, 493–558 (1909).

Der atonisch-astatische Typus der infantilen Cerebrallähmung. Dtsch. Arch. klin. Med., *98*, 216–244 (1909).

Operative Behandlung gastrischer Krisen (Med. Sekt. schles. Ges. vaterl. Kult. Breslau. 5. März 1909). Allg. med. ZentZtg. *78*, 189–190 (1909).

1910

Über die operative Behandlung spastischer Lähmungen mittels Resektion der hinteren Rückenmarkswurzeln. Berl. klin. Wschr. *47*, 1441–1444 (1910). Ref. Zbl. Neurol. Psychiat. *2*, 187–188 (1910).

Demonstration: Zwei Fälle von traumatischer Aphasie (Bresl. chir. Ges. 10. an. 1910). Berl. klin. Wschr. *47*, 313 (1910).

Ueber die Störungen in der Fixation des Beckens und Knies bei Nervenkrankheiten (9. Kongr. dtsch. Ges. orthop. Chir. Berlin. 1910). Berl. klin. Wschr. *47*, 807 (1910).

Discussion, ECKERT: Fall von Tabes mit hinterer Wurzeldurchschneidung (Ges. Charité Aerz. 3. März 1910). Berl. klin. Wschr. *47*, 1079–1080 (1910).

Die Störungen in der Fixation des Knies und Beckens bei Nervenkrankheiten. Ein Beitrag zur analytischen Uebungsbehandlung und zur orthopädischen Behandlung der Gehstörung bei Nervenkrankheiten. Z. orthop. Chir. *27*, 221–251 (1910).

1911

Diskussion, KRAMER: Zur Differentialdiagnose der Tabes dorsalis und Lues spinalis (Bresl. psychiat.-neurol. Verein. 28. Nov. 1910). Berl. klin. Wschr. *48*, 43 (1911).

Diskussion, FREUND, C. S.: Hysterischer Blepharospasmus (Bresl. psychiat.-neurol. Verein. 28. Nov. 1910). Berl. klin. Wschr. *48*, 44 (1911).

Traumatische Rückenmarksaffektionen (Bresl. psychiat.-neurol. Verein. 28. Nov. 1910). Berl. klin. Wschr. *48*, 45 (1911).

Behandlung progressiver Paralyse (Med. Sekt. schles. Ges. vaterl. Kult. Breslau. 9. Dez. 1910). Berl. klin. Wschr. *48*, 144–145 (1911).

Demonstration: Traumatische Rückenmarksaffektionen etc. (Bresl. chir. Ges. 9. Jan. 1911). Berl. klin. Wschr. *48*, 404 (1911).

Fall von gastrischen Krisen mit hinterer Wurzel-Resektion (Bresl. chir. Ges. Mai 1911). Berl. klin. Wschr. *48*, 1156 (1911).

Diskussion, KÜTTNER: Kompression des Atmungscentrums. Trepanation. Heilung (Bresl. chir. Ges. 10. Juli 1911). Berl. klin. Wschr. *48*, 1665 (1911).

Hämatomyelien (Med. Sekt. schles. Ges. vaterl. Kult. Breslau. 20. Okt. 1911). Berl. klin. Wschr. *48*, 2182 (1911).

Diskussion, HORSLEY, V.: Operative versus expectant treatment in diseases of the nervous system (4. Jahresversamml. Ges. dtsch. Nervenärz. Berlin. 6–8. Oct. 1910). Dtsch. Z. Nervenheilk. *41*, 97 (1911).

Über die Beeinflussung spastischer Lähmungen durch die Resektion hinterer Rückenmarkswurzeln (4. Jahresversamml. Ges. dtsch. Nervenärz. Berlin. 1910). Dtsch. Z. Nervenheilk. *41*, 146–169, 229–231 (1911). Ref. Zbl. ges. Neurol. Psychiat. *2*, 554–556 (1910).

Die operative Behandlung gastrischer Krisen durch Resektion hinterer Dorsalwurzeln. Ther. d. Gegenw. *52*, 337–347 (1911). Ref. Zbl. ges. Neurol. Psychiat. *5*, 83 (1912).

Ueber die operative Behandlung spastischer Lähmungen mittels Resektion hinterer Rückenmarkswurzeln. Ther. d. Gegenw. *52*, 13–18 (1911). Ref. Berl. klin. Wschr. *48*, 395 (1911).

Die Behandlung spastischer Lähmungen durch Resektion hinterer Rückenmarkswurzeln. Ergebn. Chir. Orthop. *2*, 174–209 (1911).

Diskussion, GULEKE, GÜMBEL: Erfahrungen mit der Foerster'schen Operation. Verh. dtsch. Ges. Chir. *40*, 333–334 (1911).

Bericht über luetische Affectionen des Zentralnervensystems (5. Jahresversamml. Ges. dtsch. Nervenärzte Frankfurt am Main. 1911). Verh. Ges. dtsch. Nervenärzte, 161–167 (1911).

Resection of the posterior spinal nerve-roots in the treatment of gastric crises and spastic paralysis. Proc. R. Soc. Med. *4*, 226–246 (1910–1911).

1912

Demonstration: Tuberkulöse Affektionen des Centralnervensystems, etc. (Bresl. psychiat.-neurol. Verein. 4. Dez. 1911). Berl. klin. Wschr. *49*, 184–187 (1912).

Demonstration: Cysticerkenmeningitis, etc. (Bresl. chir. Ges. 11. Dez. 1911). Berl. klin. Wschr. *49*, 279 bis 280 (1912).

Diskussion, KÜTTNER: Doppelseitige Vagotomie wegen gastrischer Krisen (Bresl. chir. Ges. 22. Jan. 1912). Berl. klin. Wschr. *49*, 570–571 (1912).

Dauerresultate der operativen Behandlung der Little'schen Krankheit mittels Wurzelresektion (Med. Sekt. schles. Ges. vaterl. Kult. Breslau. 23. Feb. 1912). Berl. klin. Wschr. *49*, 764 (1912).

Diskussion, WOLFF: Plexuslähmung bei Wirbelsäulenfraktur (Med. Sekt. schles. vaterl. Kult. Breslau. Feb. 1912). Berl. klin. Wschr. *49*, 766 (1912).

Die histologische Untersuchung der Hirnrinde intra vitam durch Hirnpunktion bei diffusen Erkrankungen des Centralnervensystems. Berl. klin. Wschr. *49*, 973–977 (1912).

Demonstration: Hämatomyelie. Sehnenplastik, etc. (Med. Sekt. schles. Ges. vaterl. Kult. Breslau. 17. Mai 1912). Berl. klin Wschr. *49*, 1251–1252 (1912).

Diskussion zu dem Referat NONNE und zu dem Vortag BENARIO. Über die sog. Neurorezidive, deren Ätiologie, Vermeidung und therapeutische Beeinflussung (5. Jahresversamml. Ges. dtsch. Nervenärz., Frankfurt 1911). Dtsch. Z. Nervenheilk. *53*, 319–325 (1912).

Diskussion, KRAUSE, F., und OPPENHEIM, H.: Zwei Fälle von cystischer Entartung des Seitenventrikels mit Hemiplegie und Epilepsie. Heilung nach beider Eröffnung und Duraplastik. Dtsch. Z. Nervenheilk. *43*, 345–346 (1912).

Arteriosklerotische Neuritis und Radiculitis (6. Jahresversamml. Ges. dtsch. Nervenärz. Hamburg, Sept. 1912). Verh. Ges. dtsch. Nervenärz. 134–165 (1912); Dtsch. Z. Nervenheilk. *45*, 374–405 (1912). Ref. Berl. klin. Wschr. *49*, 2108 (1912).

Diskussion über STOFFELSCHE Operation (11. Kongr. dtsch. Ges. orthop. Chir. Berlin. April, 1912). Z. orthop. Chir. *30* (Beilageheft), 38–50 (1912).

Die Behandlung spastischer Lähmungen mittels Resektion hinterer Rückenmarkswurzeln (11. Kongr. dtsch. Ges. orthop. Chir. Berlin. April, 1912). Z. orthop. Chir. *30*, 269–281 (1912). Ref. Berl. klin. Wschr. *49*, 870 (1912).

Demonstration zur Differentialdiagnose der Paralyse und Pseudoparalyse. Allg. Z. Psychiat. *69*, 776 bis 779 (1912).

Die Indikationen und Erfolge der Resektion hinterer Rückenmarkswurzeln. Wien. klin. Wschr. *25*, 950 bis 954 (1912). Ref. Berl. klin. Wschr. *49*, 1388 (1912).

Indications and results of excision of the posterior spinal roots in man. Med. Rec., N. Y. *82*, 916–917 (1912).

1913

Demonstration: Fall von sogenannter Torsionsneurose, etc. (Bresl. psychiat. neurol Verein. 29. Jan. 1913). Berl. klin. Wschr. *50*, 515–517 (1913).

Das phylogenetische Moment in der spastischen Lähmung (Med. Sekt. schles. Ges. vaterl. Kult. Breslau. 24. Jan. 1913). Berl. klin. Wschr. *50*, 1217–1220, 1255–1261 (1913).

Demonstration: Motorische Apraxie, etc. Berl. klin. Wschr. *50*, 1325–1328 (1913).

Vorderseitenstrangdurchschneidung im Rückenmark zur Beseitigung von Schmerzen (Med. Sekt. schles. Ges. vaterl. Kult. Breslau. 6. Juni, 1913). Berl. klin. Wschr. *50*, 1499–1500 (1913).

Meningocerebellarer Symptomenkomplex bei fieberhaften Erkrankungen (7. Jahresversamml. Ges. dtsch. Nervenärz. Breslau. Sept. 1913). Verh. Ges. dtsch. Nervenärz. 88–89 (1913). Dtsch. Z. Nervenheilk. *50*, 88–89 (1913). Ref. Dtsch. med. Wschr. *39*, 2383 (1913).

Zur Spondylitis traumatica (7. Jahresversamml. Ges. dtsch. Nervenärz. Breslau. 1913). Verh. Ges. dtsch. Nervenärz. 217–218 (1913). Dtsch. Z. Nervenheilk. *50*, 217–218 (1913) (mit SILVERBERG).

Kinematographische Demonstration: Torsionspasmus etc. (7. Jahresversamml. Ges. dtsch. Nervenärz.) Dtsch. Z. Nervenheilk. *50*, 292–294 (1913).

Die analytische Methode der kompensatorischen Uebungsbehandlung bei der Tabes dorsalis. Dtsch. med. Wschr. *39*, 1–4, 49–55, 97–101 (1913).

Demonstration: Über phylogenetische Gesichtspunkte bei der Erklärung der spastischen Lähmungen (Bresl. med. Verein. Jan. 1913). Dtsch. med. Wschr. *39*, 628 (1913).

Zur Kenntniss der spinalen Segmentinnervation der Muskeln. Neurol. Zbl. *32*, 1202–1214 (1913).

Der meningo-zerebellare Symptomenkomplex bei fieberhaften Erkrankungen tuberkulöser Individuen. Neurol. Zbl. *32*, 1414–1421 (1913). Ref. Berl. klin. Wschr. *50*, 2296 (1913).

Übungsbehandlung bei Nervenerkrankungen mit oder ohne vorausgegangene Operationen. Z. phys. diätet. Ther. *17*, 321–333, 403–415 (1913). Ref. Berl. klin. Wschr. *50*, 1411 (1913).

Die arteriosklerotische Neuritis. Wien. med. Wschr. *63*, 313–321 (1913). Ref. Zbl. ges. Neurol. Psychiat. *7*, 53 (1913).

Demonstrationen zur Hirn- und Rückenmarkschirurgie (7. Jahresversamml. Ges. dtsch. Nervenärz. Breslau. Sept. 1913). Verh. Ges. dtsch. Nervenärz. 292–294 (1913).

Traitement opératoire des paralysies spasmodiques par la résection des racines postérieures de la moëlle épinière. Paris méd. *1*, 24–28 (1913).

Die physiologischen Grundlagen der verschiedenen Behandlungsmethoden der spastischen Lähmungen (London. 7. Aug.). 17. Int. Congr. Med., 1913 (Sect. 7A Orthopedics), 7–18.

Relations between spasticity and paralysis in spastic paralysis (London. 7. Aug.). 17. Int. Congr. Med., 1913 (Sect. 11 Neuropathology), 55–64.

Les indications et les résultats de la résection des racines postérieures. 3. Clin. Congr. Surg. N. Amer. New York. 11.–16. Nov. 1912). Lyon chir. *9*, 97–109 (1913).

On the indications and results of the excision of posterior spinal nerve roots in men. Surg. Gynec. Obstet. *16*, 463–474 (1913).

1914

The borders of the areas of anesthesia, analgesia and thermoanesthesia in lesions at different levels of the sensory tract (Amer. med. Ass. Atlantic City, N. J., 24. June 1914). Titel in: J. Amer. med. Ass. *63*, 124 (1914) (anscheinend unveröffentlicht).

Diskussion, STERTZ: Die Bedeutung der Hirnpunktion für die chirurgische Indikationsstellung (Bresl. chir. Ges. 19. Jan. 1914). Berl. klin. Wschr. *51*, 375 (1914).

Demonstration: Zur Differentialdiagnose der progressiven Paralyse, etc. (Bresl. psychiat.-neurol. Verein. 23. Feb. 1914). Berl. klin. Wschr. *51*, 765–766 (1914).

1915

Demonstration: Die Schußverletzungen der peripheren Nerven und ihre Behandlung (Med. Sekt. schles. Ges. vaterl. Kult. Breslau. Mai 1915). Berl. klin. Wschr. *52*, 823–827 (1915).

1916

Die Schußverletzungen der peripheren Nerven und ihre Behandlung (Tagung dtsch. orth. Ges. Feb. 1916). Z. orthop. Chir. *36*, 310–318 (1916). Ref. Berl. klin. Wschr. *53*, 233 (1916).

Die Topik der Sensibilitätsstörungen bei Unterbrechung der sensiblen Leitungsbahnen (8. Jahresversamml. Ges. dtsch. Nervenärz. München. 1916). Neurol. Zbl. *35*, 807–808 (1916). Ref. Berl. klin. Wschr. *53*, 1230 (1916).

Kompensatorische Übungstherapie. VOGT's Handb. Ther. Nervenkr. *1*, 267–325 (1916).

Die Therapie der Motilitätsstörungen bei den Erkrankungen des Zentralnervensystems. VOGT's Handb. Ther. Nervenkr. 2, 860–944 (1916).

1917

Fall von intramedullärem Tumor, erfolgreich operiert. Berl. klin. Wschr. *54*, 338 (1917).

Diskussion, L. MANN: Über Behandlung der hysterischen Störungen bei Kriegsverletzten durch elektrische Ströme (Med. Sekt. schles. Ges. vaterl. Kult. Breslau. 3. Nov. 1916). Berl. klin. Wschr. *54*, 45 bis 46 (1917).

Die Topik der Sensibilitätstörungen bei Unterbrechung der sensiblen Leitungsbahnen (8. Jahresversamml. Ges. dtsch. Nervenärz.). Dtsch. Z. Nervenheilk. *56*, 185–186 (1917).

1918

Die psychischen Störungen der Hirnverletzten (Dtsch. Verein. Psychiat. Würzburg. 25. Apr. 1918). Allg. Z. Psychiat. *74*, 553–562 (1918).

Diskussion, KLEIST: Isolierte Fokalparesen bei isolierter Läsion der vorderen Zentralwindung. Allg. Z. Psychiat. *74*, 581–588 (1918).

Die operative Behandlung der spastischen Lähmungen (Hemiplegie, Monoplegie, Paraplegie) bei Kopf- und Rückenmarkschüssen. Dtsch. Z. Nervenheilk. *58*, 151–215 (1918). Ref. Berl. klin. Wschr. *55*, 766 (1918).

Die Symptomatologie und Therapie der Kriegsverletzungen der peripheren Nerven (9. Jahresversamml. Ges. dtsch. Nervenärz. Bonn. 22.–29. Sept. 1917). Dtsch. Z. Nervenheilk. *59*, 32–172 (1918). Ref. Berl. klin. Wschr. *54*, 1145 (1917).

Klinische Demonstration aus der Pathologie und Therapie der Verletzungen und Erkrankungen peripherer Nerven, des Rückenmarks und Gehirns (Med. Sekt. schles. Ges. vaterl. Kult. Breslau. 22. März 1918) Berl. klin. Wschr. *55*, 960 (1918).

1919

Demonstration: Nervenpfropfung bei Poliomyelitis, etc. (Med. Sekt. schles. Ges. vaterl. Kult. Breslau. 28. Feb. 1919). Berl. klin. Wschr. *56*, 741 (1919).

1920

Zwei Fälle von Angioma racemosum venosum des Gehirns (Bresl. chir. Ges. 19. Jan. 1920). Berl. klin. Wschr. *57*, 570 (1920).

Demonstration: Verletzung des Zervikalmarks, etc. (Med. Sekt. schles. Ges. vaterl. Kult. Breslau. Feb. 1920). Berl. klin. Wschr. *57*, 717–718 (1920).

Discussion, SCHÄFFER: Ueber den Antagonismus der beiden autonomen Nervensysteme an der quergestreiften Muskulatur (Med. Sekt. schles. Ges. vaterl. Kult. Breslau. 14. Mai 1920). Berl. klin. Wschr. *57*, 1007–1008 (1920).

1921

Demonstration: Zur Rindenlokalisation der Augenbewegungen. Zur Encephalitis epidemica (Med. Sekt. schles. Ges. vaterl. Kult. Breslau. 3. Dez. 1920). Ref. Berl. klin. Wschr. *58*, 458 (1921).

Diagnostik und Therapie der Rückenmarkstumoren (Bresl. chir. Ges. 6. Dez. 1920). Berl. klin. Wschr. *58*, 818–819 (1921).

Demonstration: Extramedullärer Tumor, etc. (Med. Sekt. schles. Ges. vaterl. Kult. Breslau. 17. Juni 1921). Berl. klin. Wschr. *58*, 1056–1057 (1921).

Aussprache zu den Berichten MARBURG-CASSIRER (10. Jahresversamml. Ges. dtsch. Nervenärz. Leipzig. 1920). Verh. Ges. dtsch. Nervenärz. 38–40 (1921); Dtsch. Z. Nervenheilk. *70*, 38–40 (1921).

Zur Diagnostik und Therapie der Rückenmarkstumoren (10. Jahresversamml. Ges. dtsch. Nervenärz. Leipzig. 1920). Verh. Ges. dtsch. Nervenärz. 64–74 (1921); Dtsch. Z. Nervenheilk. *70*, 64–74 (1921).

Zur Analyse und Pathophysiologie der striären Bewegungsstörungen. Z. ges. Neurol. Psychiat. *73*, 1 bis 169 (1921). Ref. Zbl. ges. Neurol. Psychiat. *29*, 42 (1922).

1922

Demonstration: Atlanto-Occipitaltuberkulose, etc. (Med. Sekt. schles. Ges. vaterl. Kult. Breslau. 20. Jan. 1922). Klin. Wschr. *1*, 1130 (1922).

Demonstrationen (Psychiat.-neurol. Verein. Breslau. 8. Mai 1922). Klin. Wschr. *1*, 1435 (1922).

Kriegsverletzungen des Rückenmarks und der peripheren Nerven. SCHJORNIG's Handb. ärz. Erfahrungen im Weltkrieg 1914–10, *4*, 235–332 (1922).

1923

Aetiologie und initiales Krankheitsbild der akuten Kinderlähmung (17. Kongr. dtsch. Orthop. Ges. Breslau. Sept. 1922). Z. orthop. Chir. *44*, 3–4 (1923).

Die Topik der Hirnrinde in ihrer Bedeutung für die Motilität (12. Jahresversamml. Ges. dtsch. Nervenärz. Halle. Okt. 1922). Dtsch. Z. Nervenheilk. *77*, 124–139 (1923). Ref. Klin. Wschr. *2*, 227 (1923).

Schlußwort (12. Jahresversamml. Ges. dtsch. Nervenärz. Halle. Okt. 1922). Dtsch. Z. Nervenheilk. *77*, 162–163 (1923).

1924

Aussprache. MINGAZZINI: Über die Pathologie des Kleinhirns (13. Jahresversamml. Ges. dtsch. Nervenärz. Danzig. Sept. 1923). Dtsch. Z. Nervenheilk. *81*, 55–56 (1924).

Aussprache. DUSSER DE BARENNE: Experimentelle Untersuchungen über die Localisation (14. Jahresversamml. Ges. dtsch. Nervenärz.). Dtsch. Z. Nervenheilk. *83*, 300–301 (1924).

Hyperventilationsepilepsie (14. Jahresversamml. Ges. dtsch. Nervenärz. Innsbruck. 24.–27. Sept. 1924). Verh. Ges. dtsch. Nervenärz. 155–163 (1925); Dtsch. Z. Nervenheilk. *83*, 347–356 (1924). Ref. Klin. Wschr. *3*, 2269 (1924).

Demonstration: Traumatisches Aneurysma der Arteria cerebri anterior (Psychiat. neurol. Verein. Breslau. 26. Okt. 1923). Klin. Wschr. *3*, 170 (1924).

Zur operativen Behandlung der Epilepsie (Bresl. chir. Ges. 10. Nov. 1924). Klin. Wschr. *3*, 2412 (1924).

Umfrage über die periarterielle Sympathectomie. Med. Klinik *20*, 532–535 (1924).

Ein Fall mit ungewöhnlichen Augensymptomen bei Encephalitis (Verein. Augenärz. Schles. u. Posens, Breslau. 18. Mai 1924). Klin. Mbl. Augenheilk. *73*, 247–249 (1924) (mit BIELSCHOWSKY, S. [1]).

1925

Über die therapeutische Verwendbarkeit des Tetrophans. Klin. Wschr. *4*, 55–60 (1925). Ref. Zbl. ges. Neurol. Psychiat. *4*, 41–52 (1925).

Encephalographische Erfahrungen. Z. ges. Neurol. Psychiat. *94*, 512–584 (1925).

Zur Pathogenese und chirurgischen Behandlung der Epilepsie (Bresl. chir. Ges. 10. Nov. 1924). Zbl. Chir. *52*, 531–549 (1925).

Ueber die antidrome Leitung der sensiblen Nerven, pp. 145–155 in: Festschrift für Prof. G. ROSSOLIMO 1884–1924). Berlin, 1925. Ref. Zbl. ges. Neurol. Psychiat. *43*, 625 (1926).

1926

Ansprache (15. Jahresversamml. Ges. dtsch. Nervenärz. Cassel. 3. Sept. 1925). Dtsch. Z. Nervenheilk. *88*, 99–113 (1926).

Zur operativen Behandlung der Epilepsie (15. Jahresversamml. Ges. dtsch. Nervenärz. Cassel. 1925). Dtsch. Z. Nervenheilk. *89*, 137–147 (1926).

Ansprache (16. Jahresversamml. Ges. dtsch. Nervenärz. Düsseldorf. 1926). Verh. Ges. dtsch. Nervenärz. 3–14 (1926); Dtsch. Z. Nervenheilk. *94*, 3–14 (1926).

Die Pathogenese des epileptischen Krampfanfalles (16. Jahresversamml. Ges. dtsch. Nervenärz.). Verh. Ges. dtsch. Nervenärz. 15–53 (1926); Dtsch. Z. Nervenheilk. *94*, 15–53 (1926).

Schlußwort. WUTH, O.: Stoffwechselpathologie (16. Jahresversamml. Ges. dtsch. Nervenärz. Düsseldorf. 1926). Dtsch. Z. Nervenheilk. *94*, 119–122 (1926).

Ansprache (16. Jahresversamml. Ges. dtsch. Nervenärz. Düsseldorf. 1926). Verh. Ges. dtsch. Nervenärz. 131–139 (1926). Dtsch. Z. Nervenheilk. *94*, 131–135 (1926).

Aussprache. SCHWAB, O.: Über vorübergehende aphasische Störungen nach Rindenexzision usw. (16. Jahresversamml. Ges. dtsch. Nervenärz. Düsseldorf. 1926). Dtsch. Z. Nervenheilk. *94*, 182–183 (1926).

Aussprache. TATERKA, H.: Über Spontanblutungen bei Tabes dorsalis usw. (16. Jahresversamml. Ges. dtsch. Nervenärz. Düsseldorf. 1926). Dtsch. Z. Nervenheilk. *94*, 196 (1926).

Aussprache. LEWY, F. H.: Die Bedeutung der Infektion für die Neurologie (16. Jahresversamml. Ges. dtsch. Nervenärz. Düsseldorf. 1926). Dtsch. Z. Nervenheilk. *94*, 205 (1926).

Schädigung des Gehirns durch stumpfe Kopfverletzungen (12. Tagung südostdtsch. chir. Verein.). Zbl. Chir. *53*, 1192–1198 (1926).

Operative Behandlung des Torticollis spasticus (13. Tagung südostdtsch. chir. Verein. Breslau. 26. Juli 1926). Zbl. Chir *53*, 2804–2805 (1926).

Demonstration: Fibrosarkom der oberen Halswirbel mit Halsmarkkompression, etc. (Psychiat.-neurol. Verein. Breslau. 17. Dez. 1925). Klin. Wschr. *5*, 432–433 (1926).

Demonstration: Plötzliche Kompression des Opticus, etc. (Verein. südostdtsch. Psychiat. Neurol. Breslau. 17. Mai 1926). Klin. Wschr. *5*, 1896, 1897 (1926).

Methoden der Dermatombestimmung beim Menschen (Verein. südostdtsch. Neurol. Psychiat. Breslau. 27. bis 28. März 1926). Arch. Psychiat. Nervenkr. *77*, 652–658 (1926).

1927

Die Leitungsbahnen des Schmerzgefühls und die chirurgische Behandlung der Schmerzzustände. Bruns Beitr. klin. Chir., 1927, 360 pp. (Sonderbd.). Ref. Klin. Wschr. *6*, 470–471 (1927).

Demonstration: Methoden der Bestimmung des Höhensitzes spinaler Transversalläsionen, etc. (Verein. südostdtsch. Psychiat. Neurol. 5. Mai 1927). Klin. Wschr. *6*, 1825 (1927).

Über die Vorderseitenstrangdurchschneidung (2. Jahresversamml. Verein. südostdtsch. Psychiat. Neurol. Breslau. 5.–6. März 1927). Arch Psychiat. Nervenkr. *81*, 707–717 (1927).

Schlaffe und spastische Lähmung. Handb. norm. path. Physiol. *10*, 893–972 (1927).

1928

Ansprache (17. Jahresversamml. Ges. dtsch. Nervenärz. Wien. Sept. 1927). Verh. Ges. dtsch. Nervenärz. 4–27 (1928); Dtsch. Z. Nervenheilk. *101*, 88–110 (1928).

Ansprache (18. Jahresversamml. Ges. dtsch. Nervenärz. Hamburg. 1928). Dtsch. Z. Nervenheilk. *106*, 112 bis 136 (1928).

Zur Pupillarinnervation. Dtsch. Z. Nervenheilk. *106*, 311–313 (1928).

Über die Vasodilatatoren in den peripheren Nerven und hinteren Rückenmarkswurzeln beim Menschen. Dtsch. Z. Nervenheilk. *107*, 41–56 (1928).

Ein Fall von Vierhügeltumor durch Operation entfernt (3. Jahresversamml. südostdtsch. Psychiat. Neurol. Breslau. 26. Feb. 1928). Arch. Psychiat. Nervenkr. *84*, 515–516 (1928).

Das operative Vorgehen bei Tumoren der Vierhügelgegend (Festschr. Wagner-Jauregg). Wien. klin. Wschr. *41*, 986–990 (1928). Ref. Klin. Wschr. *7*, 2267 (1928).

1929

Über die Beziehung des vegetativen Nervensystems zur Sensibilität. Jb. schles. Ges. vaterl. Kult. *102* (Med. Sekt.), 1–3 (1929); Med. Klinik *25*, 519–520 (1929). Zusfssg. Klin. Wschr. *8*, 713–714 (1929) (mit H. ALTENBURGER).

Über die Nachbarschaftssymptome der Hypophysentumoren (2. Südostdtsch. Ärztetagung. Prag. 23. bis 24. Feb. 1929). Med. Klinik *25*, 925–926 (1929).

Demonstration: Spätoperation nach Schußverletzungen peripherer Nerven, etc. (Bresl. chir. Ges. 16. Jan. 1929). Klin. Wschr. *8*, 522 (1929); Zbl. Chir. *56*, 891–895 (1929).

Über die Beziehungen des vegetativen Nervensystems zur Sensibilität. Z. ges. Neurol. Psychiat. *121*, 139 bis 185 (1929). Ref. Klin. Wschr. *8*, 2256 (1929); Zbl. ges. Neurol. Psychiat. *55*, 29 (1930) (mit H. ALTENBURGER und F. W. KROLL).

Beiträge zur Pathophysiologie der Sehbahn und der Sehsphäre. J. Psychol. Neurol. Lpz. *39*, 463–485 (1929). Ref. Zbl. ges. Neurol. Psychiat. *56*, 60 (1930).

Ansprache (19. Jahresversamml. Ges. dtsch. Nervenärz. Würzburg. Sept. 1929. Verh. Ges. dtsch. Nervenärz. 4–16 (1929); Dtsch. Z. Nervenheilk. *110*, 208–220 (1929).

Encephalographische Erfahrungen (4. Jahresversamml. südostdtsch. Psychiat. Neurol. Breslau. 2.–3. März 1929). Arch. Psychiat. Nervenkr. *88*, 462–467 (1929).

Über die Beziehungen zwischen vegetativem Nervensystem und Sensibilität (Schles. Ges. vaterl. Kult. Breslau. 1. Feb. 1929). Dtsch med. Wschr. *55*, 728 (1929) (mit H. ALTENBURGER).

Torticollis spasticus (23. Kongr. dtsch. Orthop. Ges. Prag. Sept. 1928). Z. orthop. Chir. *51*, 144–168 (1929) (Beilageheft).

Spezielle Anatomie und Physiologie der peripheren Nerven. Handb. Neurol. (LEWANDOWSKY) *3*, 785–974 (1929) (Ergänzbd.).

Die Symptomatologie der Schußverletzungen der peripheren Nerven. Handb. Neurol. (LEWANDOWSKY) *3*, 975–1508 (1929) (Ergänzbd.).

Die Therapie der Schußverletzungen der peripheren Nerven. Handb. Neurol. (LEWANDOWSKY) *3*, 1509 bis 1720 (1929) (Ergänzbd.).

Die traumatischen Läsionen des Rückenmarkes auf Grund der Kriegserfahrungen (Der Mechanismus ihres Zustandekommens und die pathologisch-anatomischen Veränderungen). Handb. Neurol. (LEWANDOWSKY) *3*, 1721–1927 (1929) (Ergänzbd.).

1930

Ansprache (20. Jahresversamml. Ges. dtsch. Nervenärz. Dresden 1930). Dtsch. Z. Nervenheilk. *115*, 147 bis 159 (1930); Verh. Ges. dtsch. Nervenärz., 3–15 (1931).

Klinische Beiträge: I. Restitution der Motilität. II. Restitution der Sensibilität (20. Jahresversamml. Ges. dtsch. Nervenärz. Dresden. Sept. 1930). Dtsch. Z. Nervenheilk. *115*, 248–295, 296–314 (1930); Verh. Ges. dtsch. Nervenärz., 104–170 (1931).

Schlußwort. GOLDSTEIN: Restitution bei Schädigungen der Hirnrinde (20. Jahresversamml. Ges. dtsch. Nervenärz. Dresden. Sept. 1930). Dtsch. Z. Nervenheilk. *116*, 42–43 (1930).

Der Narbenzug am und im Gehirn bei traumatischer Epilepsie in seiner Bedeutung für das Zustandekommen der Anfälle und für die therapeutische Bekämpfung derselben. Z. ges. Neurol. Psychiat. *125*, 475–572 (1930) (mit W. PENFIELD).

Demonstration mehrerer Fälle von raumbeengenden Prozessen der hinteren Schädelgrube (Verein. südostdtsch. Psychiat. Neurol. Breslau. 20. Jan. 1930). Klin. Wschr. *9*, 1603–1604 (1930).

Beitrag zur Behandlung spondylitischer Prozesse im Bereich des Atlas und Epistropheus. Fixierung des Kopfes und der Halswirbelsäule durch Implantation eines Fibulastückes zwischen Vertebra prominens und Okziput. Festschrift S. E. HENSCHEN. J. Psychol. Neurol., Lpz. *40*, 215–224 (1930). Ref. Zbl. ges. Neurol. Psychiat *57*, 335 (1930).

Über die efferenten Fasern in den hinteren Wurzeln (5. Jahresversamml. Verein südostdtsch. Psychiat. Neurol. Breslau. März 1930). Arch. Psychiat. Nervenkr. *91*, 474–475 (1930) (mit O. GAGEL).

Beitrag zum Werte fixierender orthopädischer Operationen bei Nervenkrankheiten. Festschrift P. HAGLUND. Acta chir. scand. *67*, 351–376 (1930).

The structural basis of traumatic epilepsy and results of radical operation. Brain *53*, 99–119 (1930) (with W. PENFIELD).

1931

Über das Phantomglied. Med. Klinik *27*, 497–500 (1931). Ref. Zbl. ges. Neurol. Psychiat. *61*, 72 (1931).

Demonstration (Schles. Ges. vaterl. Kult. Breslau. 10. Juli 1931). Med. Klin. *27*, 1367–1368 (1931).

Demonstration: Hintere Wurzeldurchschneidung bei Schmerzzuständen, etc. (Verein. südostdtsch. Neurol. Psychiat. Breslau. 23. Juli 1930). Klin. Wschr. *10*, 329 (1931).

Ein Fall von sogenanntem Gliom des Nervus opticus- Spongioblastoma multiforme ganglioides. Z. ges. Neurol. Psychiat. *136*, 335–366 (1931). Ref. Zbl. ges. Neurol. Psychiat. *62*, 589 (1932) (mit O. GAGEL).

La ventriculographie dans les tumeurs du mésocéphale, du diencéphale et dans les pseudotumeurs. Rev. neurol. *56*, 369–370 (1931).

Le processus opératoire dans les tumeurs de la région quadrigéminale. Rev. neurol. *56*, 481 (1931).

The results of electrical stimulation of the cortex cerebri in man, their relations to architectonic structure, to the results of experimental physiology and to clinical symptomatology. Abstr. under title: The cerebral cortex in man. Lancet *2*, 309–312 (1931).

Surgical treatment of neurogenic contractures (20th Ann. Clin. Congr. Amer. Coll. Surg. Philadelphia. 13–17th Oct. 1930). Surg. Gynec. Obstet. *52*, 360–366 (1931). Ref. Zbl. ges. Neurol. Psychiat. *60*, 794 (1931).

Rede (1st Int. Neurol Congr. Bern. 3. Sept. 1931). Publ. in: Bull. Hist. Med. *8*, 332–354 (1940).

1932

Die Vorderseitenstrangdurchschneidung beim Menschen. Eine klinisch-patho-physiologisch-anatomische Studie. Z. ges. Neurol. Psychiat. *138*, 1–92 (1932). Ref. Klin. Wschr. *11*, 1561 (1932) (mit O. GAGEL).

Ein Fall von Recklinghausenscher Krankheit mit fünf nebeneinander bestehenden verschiedenartigen Tumorbildungen. Z. ges. Neurol. Psychiat. *138*, 339–360 (1932). Ref. Klin. Wschr. *12*, 400 (1933) (mit O. GAGEL).

Über die Beziehung von Vorstellung und Wahrnehmung bei Schädigung afferenter Leitungsbahnen. Z. ges. Neurol. Psychiat. *139*, 658–693 (1932). Ref. Klin. Wschr. *12*, 163 (1933) (mit M. LOEWI).

Ein Fall von Gangliocytom der Oblongata. Z. ges. Neurol. Psychiat. *141*, 797–823 (1932) (mit O. GAGEL).

Ein Fall von Gangliogliom der Rautengrube. Z. ges. Neurol. Psychiat. *142*, 507–518 (1932). Ref. Zbl. ges. Neurol. Psychiat. *67*, 448 (1933) (mit O. GAGEL).

Die Bedeutung der Ventriculographie für die Diagnose der Tumoren des Mittel- und Zwischenhirns und für die Differentialdiagnose zwischen Tumor cerebri und Pseudotumor cerebri (Int. Neurol. Kongr. Bern. 1931). Zbl. ges. Neurol. Psychiat. *61*, 441–445 (1932).

Über das operative Vorgehen bei Tumoren der Vierhügelgegend (Int. Neurol. Kongr. Bern. Sept. 1931). Zbl. ges. Neurol. Psychiat. *61*, 457–459 (1932).

Demonstration: Meningiome, etc. (Verein. südostdtsch. Neurol. Psychiat. Breslau. 2. Mai 1932). Klin. Wschr. *11*, 1892–1893 (1932).

1933

Ein Fall von Ganglioneuroma amyelinicum des Hirnstammes. Z. ges. Neurol. Psychiat. *143*, 635–650 (1933) (mit A. J. McLean und O. Gagel).

Über afferente Nervenfasern in den vorderen Wurzeln. Z. ges. Neurol. Psychiat. *144*, 313–324 (1933). Ref. Klin. Wschr. *12*, 1887 (1933) (mit O. Gagel).

Ein Fall von Gangliogliom der Regio hypothalamica. Z. ges. Neurol. Psychiat. *145*, 17–28 (1933). Ref. klin. Wschr. *12*, 1586 (1933) (mit A. J. McLean und O. Gagel).

Ein Fall von Gangliogliom des Bodens des dritten Ventrikels. Z. ges. Neurol. Psychiat. *145*, 29–37 (1933). Ref. Klin. Wschr. *12*, 1586 (1933) (mit O. Gagel).

I. Zur elektrophysiologischen Analyse der Sehnen- und Knochenphänomene und der Dehnungsreflexe. Neurol. Psychiat. *146*, 641–660 (1933) (mit H. Altenburger).

II. Die Dehnungsreflexe bei Gesunden. Ibid. *147*, 169–183. Ref. Klin. Wschr. *13*, 312 (1934) (mit H. Altenburger).

III. Die Sehnen- und Knochenphänomene beim Pyramidenbahnsyndrom. Ibid. *147*, 779–790. Ref. Klin. Wschr. *13*, 312 (1934) (mit H. Altenburger).

IV. Die Dehnungs- und Annäherungsreflexe beim Pyramidenbahnsystem. Ibid. *148*, 655–669 (mit H. Altenburger).

V. Die Reflexsynergien beim Pyramidenbahnsyndrom. Ibid. *149*, 409–418 (mit H. Altenburger).

Ein Fall von Gangliocytoma dysplasticum des Kleinhirns. Z. ges. Neurol. Psychiat. *146*, 792–803 (1933) (mit O. Gagel).

Nucleare Lähmungen bei anämischer funikulärer Spinalerkrankung und ihre Behandlung. Z. ges. Neurol. Psychiat. *147*, 161–168 (1933). Ref. Klin. Wschr. *12*, 1784 (1933) (mit G. Hofheinz und L. Guttmann).

Ein Fall von Ganglienzellgeschwulst des Hirnstammes (N. caudatus). Z. ges. Neurol. Psychiat. *147*, 713 bis 745 (1933). Ref. Klin. Wschr. *13*, 1480 (1934) (mit O. Gagel und A. J. McLean).

Ein Fall von Ependymcyste des III. Ventrikels. Ein Beitrag zur Frage der Beziehungen psychischer Störungen zum Hirnstamm. Z. ges. Neurol. Psychiat. *149*, 312–344 (1933). Ref. Klin. Wschr. *13*, 1292 (1934) (mit O. Gagel).

Ansprache zur Eröffnung der 21. Jahresversamml. Ges. dtsch. Nervenärz. Wiesbaden. 22.–24. Sept. 1932. Verh. Ges. dtsch. Nervenärz., 3–12 (1933); Dtsch. Z. Nervenheilk. *129*, 175–184 (1933).

Ansprache bei Überreichung der Erb-Denkmünze W. Spielmeyer (21. Jahresversamml. Gesch. dtsch. Nervenärz. Wiesbaden. 23. Sept. 1932). Dtsch. Z. Nervenheilk. *130*, 1–3 (1933).

Demonstration: Messerstichverletzungen des Rückenmarks, etc. (Verein. südostdtsch. Neurol. Psychiat. Breslau. 26. Nov. 1932). Klin. Wschr. *12*, 923 (1933).

Cerebrale Komplikationen bei Thromboangiitis obliterans. Arch. Psychiat. Nervenkr. *100*, 506–515 (1933). Ref. Klin. Wschr. *13*, 151 (1934) (mit L. Guttmann).

Veränderungen an den Endösen im Rückenmark des Affen nach Hinterwurzeldurchschneidung. Z. ges. Anat. I. Z. Anat. EntwGesch. *101*, 553–565 (1933) (mit O. GAGEL und D. SHEEHAN).

Symptomatische Eingriffe am Nervensystem, insbesondere solche der Schmerzbekämpfung. Münch. med. Wschr. *80*, 83–87 (1933). Ref. Zbl. ges. Neurol. Psychiat. *67*, 587 (1933).

Ueber einen Fall von Stichverletzung des Rückenmarkes, ein Beitrag zur Lehre von der Funktion der medullären sensiblen Leitungsbahnen, insbesondere der Hinterstränge. Pp. 213–228 in: Volume Jubilaire en l'Honneur du G. MARINESCO. Bucharest. E. Marvan (1933).

The dermatomes in man (Schorstein Lecture, London, 1932). Brain *56*, 1–39 (1933).

Mobile spasm of the neck muscles and its pathological basis. J. Comp. Neurol. *58*, 725–735 (1933).

1934

Zur Physiologie und Pathophysiologie der Sehnen- und Knochenphänomene und der Dehnungs- und Adaptionsreflexe. VI. Die Sehnen und Knochenphänomene beim Pallidumsyndrom. Z. ges. Neurol. Psychiat. *150*, 163–171 (1934) (mit H. ALTENBURGER).

VII. Die Dehnungs- und Annäherungsreflexe beim Pallidiumsyndrom. Ibid., 588–596 (mit H. ALTENBURGER).

Ein Fall von Ependymoma polycysticum des Kleinhirns. Z. ges. Neurol. Psychiat. *150*, 515–527 (1934). Ref. Klin. Wschr. *14*, 514 (1935) (mit O. GAGEL).

Zentrale diffuse Schwannose bei Recklinghausenscher Krankheit. Z. ges. Neurol. Psychiat. *151*, 1–16 (1934). Ref. Klin. Wschr. *14*, 618 (1935) (mit O. GAGEL).

Die Diagnostik und Behandlung der Geschwülste des Großhirns. Klin. Wschr. *13*, 1737–1742 (1934).

Die tigrolytische Reaktion der Ganglienzelle. Z. mikr.-anat. Forsch. *36*, 567–575 (1934) (mit O. GAGEL).

Die operative Behandlung der Schußverletzungen der peripheren Nerven. Münch. med. Wschr. *81*, 1183 bis 1187 (1934).

Über die Bedeutung und Reichweite des Lokalisationsprinzips im Nervensystem (46. Kongr. Wiesbaden 1934). Verh. dtsch. Ges. inn. Med. *46*, 117–211 (1934). Ref. Klin. Wschr. *13*, 678 (1934).

1935

Elektrobiologische Vorgänge an der menschlichen Hirnrinde (22. Jahresversamml. Ges. dtsch. Nervenärz. München. 1934). Dtsch. Z. Nervenheilk. *135*, 277–286 (1935) (mit H. ALTENBURGER).

Klinik und Pathohistologie der intramedullären Rückenmarkstumoren. Dtsch. Z. Nervenheilk. *136*, 239 (1935) (mit O. GAGEL).

Über Störungen der Thermoregulation bei Erkrankungen des Gehirns und Rückenmarks und bei Eingriffen am Zentralnervensystem. Jb. Psychiat. Neurol. *52*, 1–14 (1935).

Der Schmerz und seine operative Bekämpfung. Nova Acta Leop. Carol. *3* (n. F.), 1–60 (1935).

1936

Zum Geleit. Zbl. Neurochir. *1*, 2–3 (1936).

Das Ependymom des Filum terminale. Zbl. Neurochir. *1*, 5–18 (1936) (mit O. GAGEL).

Zur Physiologie und Pathophysiologie der Sehnen- und Knochenphänomene und der Dehnungsreflexe. VIII. Die Sehnen- und Knochenphänomene und der Dehnungsreflex beim Cerebellarsyndrom. Z. ges. Neurol. Psychiat. *156*, 479–483 (1936) (mit H. ALTENBURGER).

Über die Anatomie, Physiologie und Pathologie der Pupillarinnervation (48. Kongr. Wiesbaden. 1936). Verh. dtsch. Ges. inn. Med. *48*, 386–398 (1936) (mit O. GAGEL und W. MAHONEY).

Symptomatologie der Erkrankungen des Rückenmarks und seiner Wurzeln. BUMKE u. FOERSTERs Handb. Neurol. *5*, 1–403 (1936).

Motorische Felder und Bahnen. BUMKE u. FOERSTERs Handb. Neurol. Springer-Verlag *6*, 1–357 (1936).

Sensible corticale Felder. BUMKE u. FOERSTERs Handb. Neurol. Springer-Verlag *6*, 358–448 (1936).

Übungstherapie. BUMKE u. FOERSTERs Handb. Neurol. Springer-Verlag *8*, 316–414 (1936).

The motor cortex in man in the light of Hughlings Jackson's doctrines. Brain *59*, 135–159 (1936).

A contribution to the study of gliomas of the spinal cord with special reference to their operability. Pp. 9–67 in: Jubilee Vol. for Davidenkov. Leningrad, State Inst. for Publ. Biol. and Med. Literature, 1936) (mit P. BAILEY).

1937

Die Tumoren der Brücke. I. Ein Fall von Astrocytom der Brücke. Z. ges. Neurol. Psychiat. *157*, 136–146 (1937) (mit P. C. BUCY [1], O. GAGEL [3] und W. MAHONEY [4]).

Vegetative Regulationen (49. Kongr. dtsch. Ges. inn. Med. Wiesbaden. 1937). Verh. dtsch. Ges. inn. Med. *49*, 165–187 (1937) (mit O. GAGEL und W. MAHONEY).

Spezielle Physiologie und spezielle funktionelle Pathologie der quergestreiften Muskeln. BUMKE u. FOERSTERs Handb. Neurol. Springer-Verlag *3*, 1–639 (1937).

1938

Ein Fall von Hämatomyelie des oberen Halsmarkes. Zbl. Neurochir. *3*, 321–329 (1938).

Aussprache: Über das Chiasmasyndrom. Dtsch. med. Wschr. *64*, 188–190 (1938).

Die Hirntumoren und ihre moderne Diagnostik und Therapie. Neue dtsch. Klinik *16*, 44–64 (1938). (Ergänzungsbd. 6).

Über die Wechselbeziehungen von Herdsymptomen und Allgemeinsymptomen beim Hirntumor. Verh. dtsch. Ges. inn. Med. *50*, 458–485 (1938).

1939

Ein Fall von Agenesie des Corpus callosum verbunden mit einem Diverticulum paraphysarium des Ventriculus tertius. Z. ges. Neurol. Psychiat. *164*, 380–391 (1939).

Das umschriebene Arachnoidealsarkom des Kleinhirns. Z. ges. Neurol. Psychiat. *164*, 565–580 (1939). Zusfssg. Arch Neurol. Psychiat., Chicago, *42*, 1147–1148 (1939) (mit O. GAGEL).

Die Astrocytome der Oblongata, Brücke und des Mittelhirns. Z. ges. Neurol. Psychiat. *166*, 497–528 (1939) (mit O. GAGEL).

Überreichung der Erb-Denkmünze an ERNST RÜDIN und HEINRICH PETTE. Z. ges. Neurol. Psychiat. *167*, 6–8 (1939).

Operativ-experimentelle Erfahrungen beim Menschen über den Einfluß des Nervensystems auf den Kreislauf. Verh. dtsch. Ges. inn. Med. *51*, 253–275 (1939). Z. ges. Neurol. Psychiat. *167*, 439–461 (1939).

Thyreogene intrarachideale Geschwülste. Zbl. Neurochir. *4*, 198–214 (1939).

HARVEY CUSHING. Zbl. Neurochir. *4*, 195–197 (1939).

Die topische Tumordiagnostik, Herdsymptome des Hirntumores. Neue dtsch. Klinik *16*, 260–299 (1939) (Ergänzungsbd. 6).

Tumorartdiagnostik, Röntgendiagnostik, Hirnpunktion, Lumbalpunktion. Neue dtsch. Klinik *16*, 419 bis 468 (1939) (Ergänzungsbd. 6).

Die encephalen Tumoren des verlängerten Markes, der Brücke und des Mittelhirns. Arch. Psychiat. Nervenkr. *110*, (1939) (mit O. GAGEL und W. MAHONEY).

1940

Die encephalen Tumoren der Oblongata, Pons und des Mesencephalons. III. Z. ges. Neurol. Psychiat. *168*, 295–331 (1940) (mit O. GAGEL) IV. Ibid., 492–518.

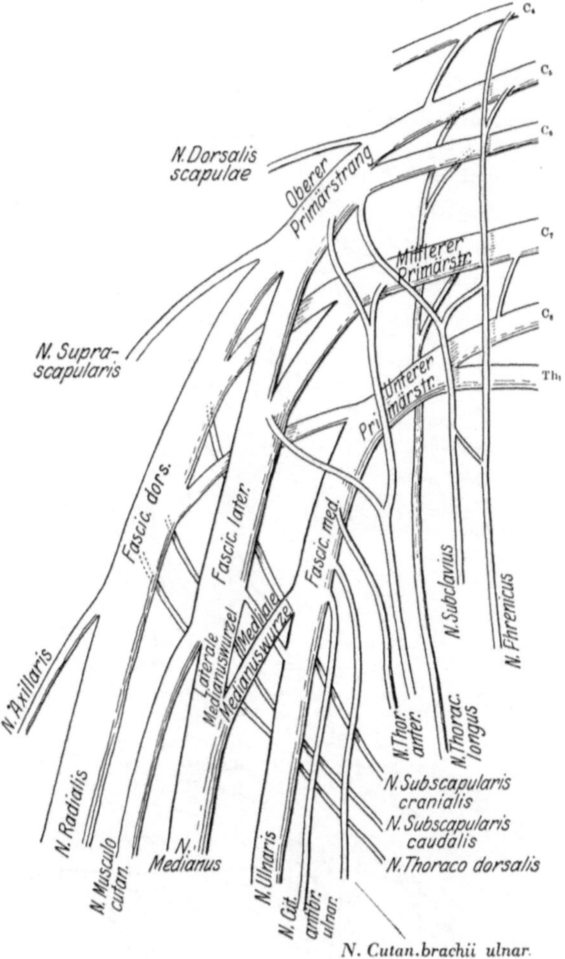

Fig. 1. Schematic drawing of the brachial plexus

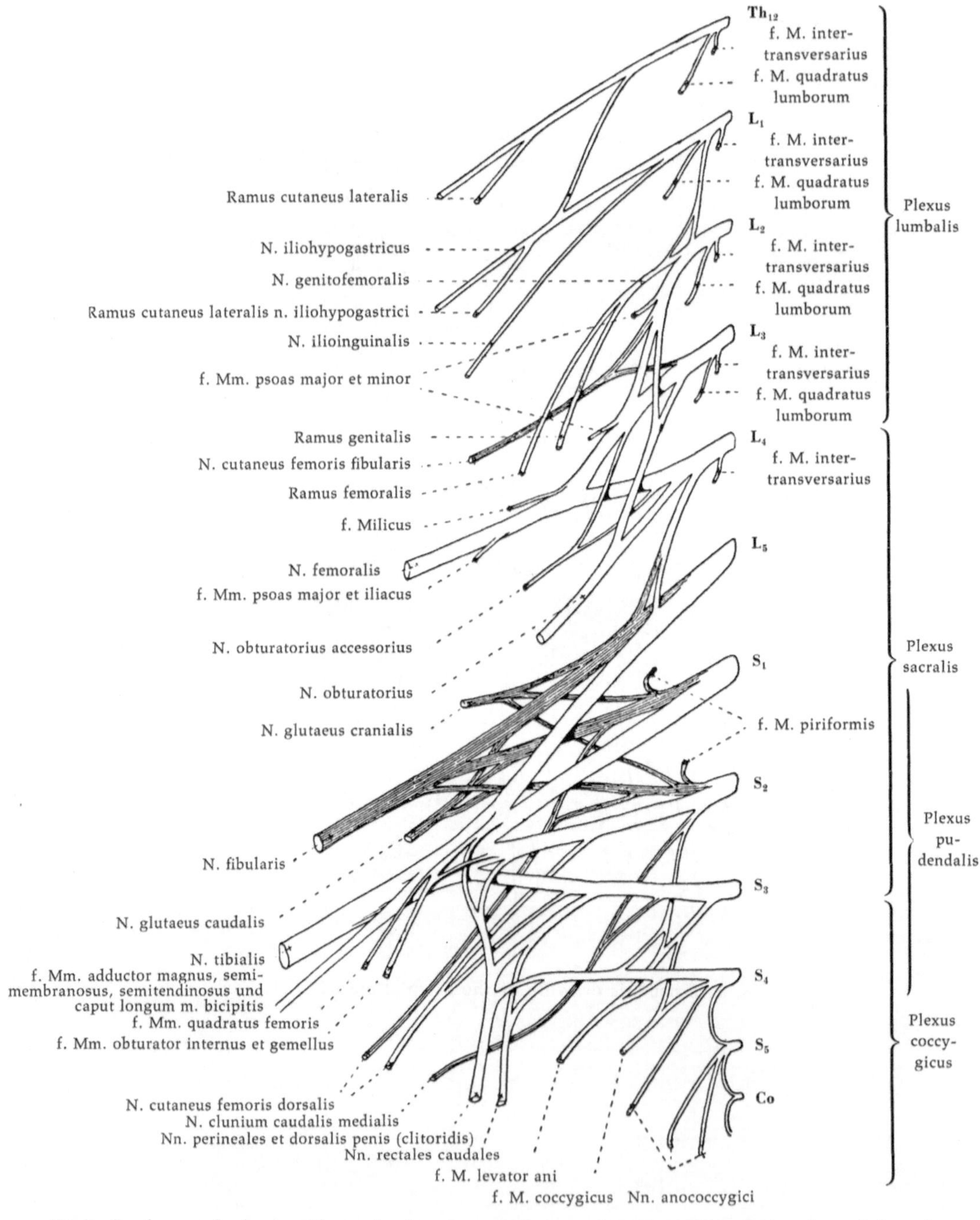

. 2. Right lumbosacral plexus. Schematic drawing of the anterior view. The dark coloured branches ne from the dorsal half of the plexus

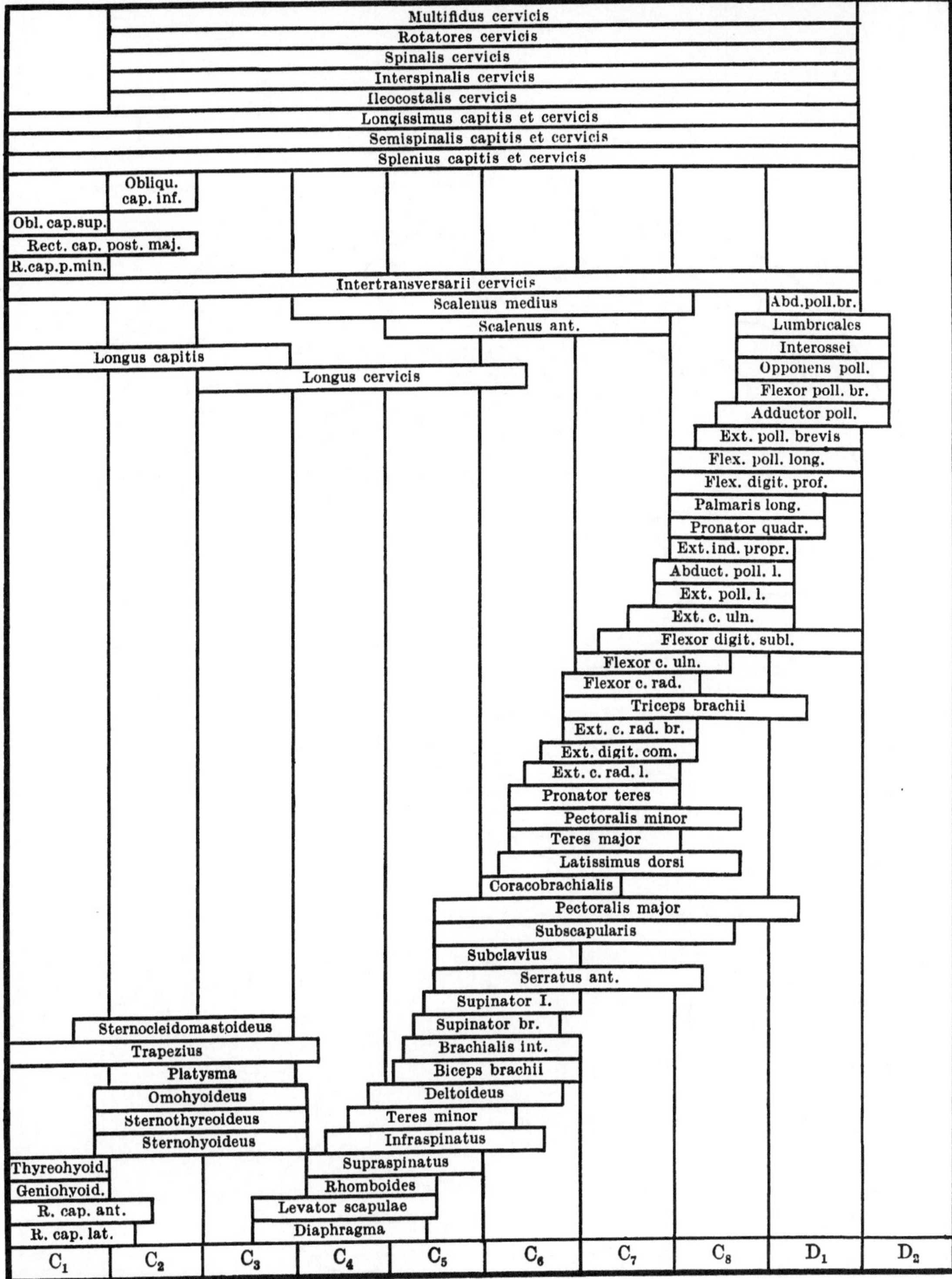

Fig. 3. Segmental Columns of the Anterior Motor Horn Cells Innervating Muscles of the Neck and Upper Extremity

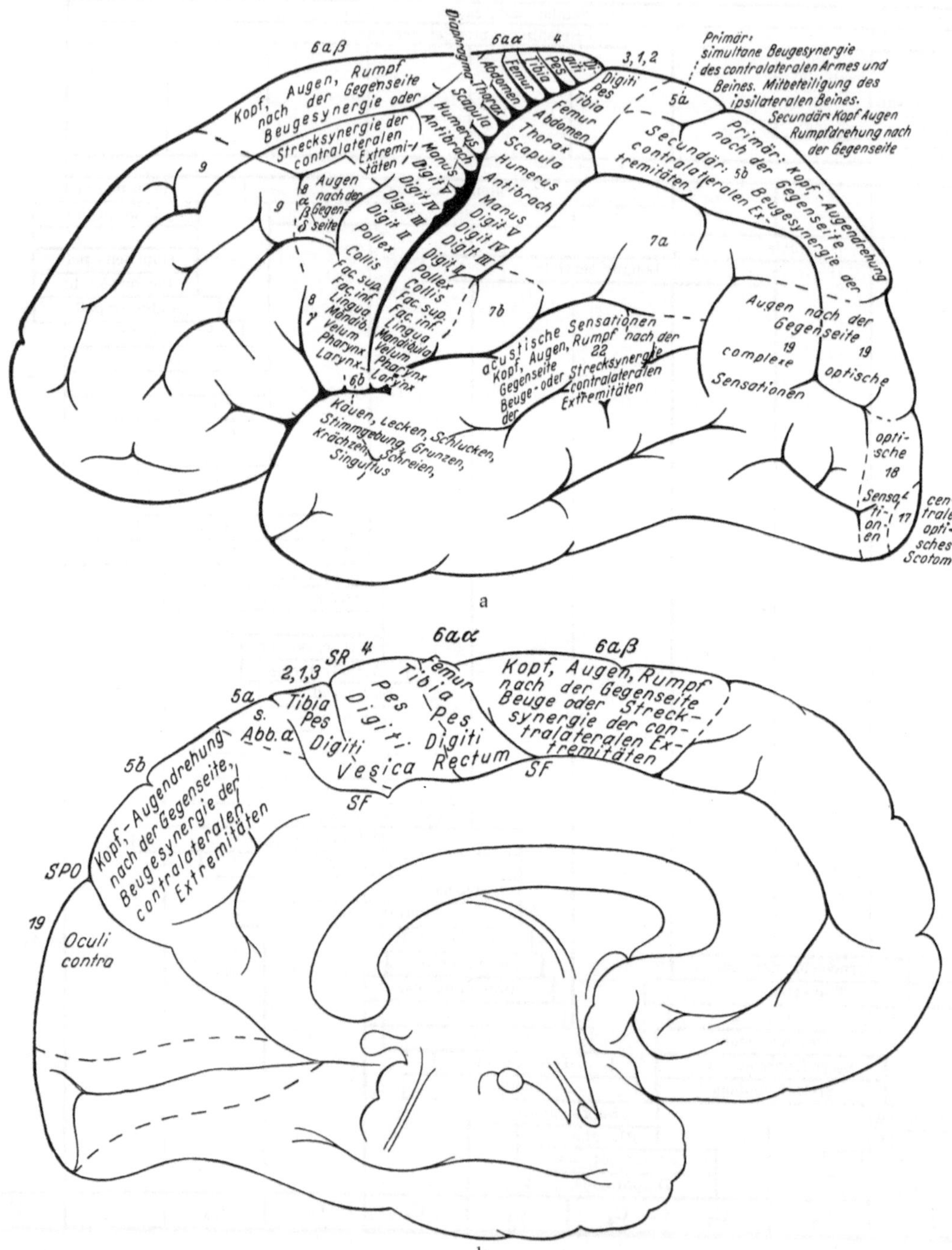

Fig. 4. The effects elicited by electric stimulation of the motor areas

Fig. 5. OTFRID FOERSTER in the last decade of his life

6. "In 1932 . . . the Rockefeller Foundation built a modern institute which was supported by the city ꞏreslau, the Province of Silesia and the University"

Fig. 7. "The close relationship to Anglo-Saxons stemmed from genuine congeniality and also from a shared scientific methodology"

above: ALFRED ADSON (Mayo Clinic), OTFRID FOERSTER, MAX PEET and EDGAR KAHN (both University of Michigan)

below: O. GAGEL, O. FOERSTER, J. F. FULTON (Yale University), M. KENNARD (Yale University)

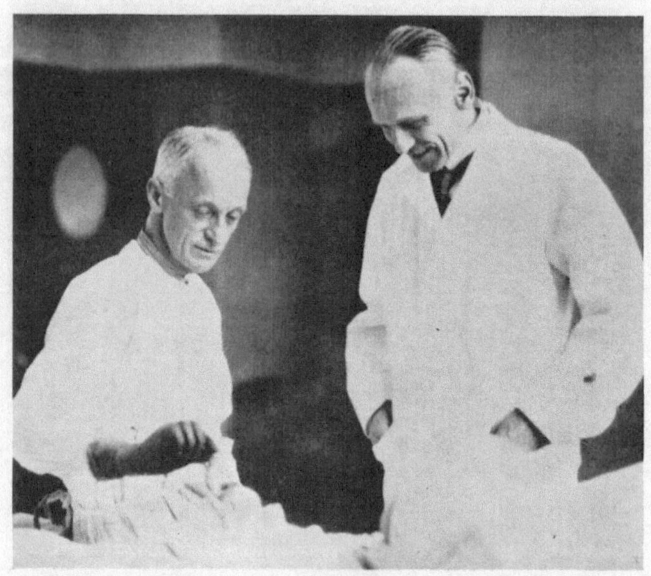

8. "... The sojourn in 1930 in Cushing's Clinic, where he was named Surgeon-in-Chief pro tempore, not change his operative techniques ..."

Fig. 9. "This tumor classification, executed in the minutest detail, is the life work of HARVEY CUSHING, great not only from a pathological point of view but above all from the standpoint of practical therapeutics"

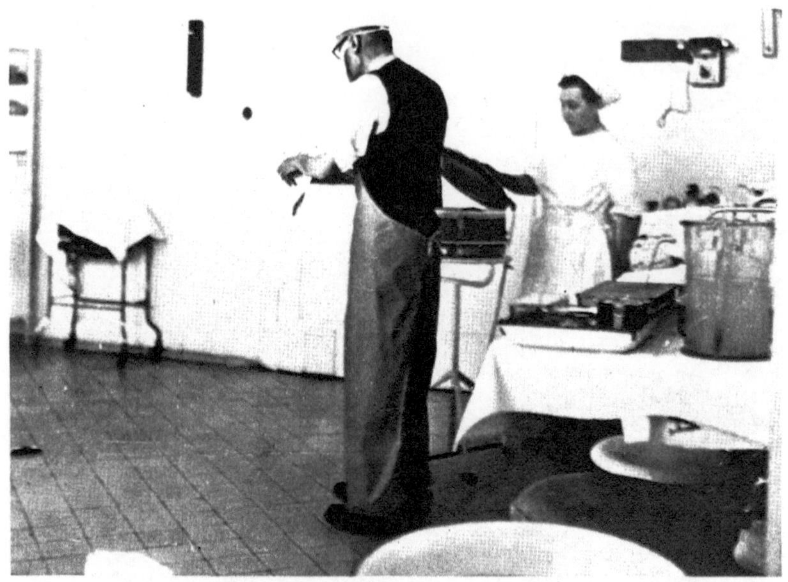

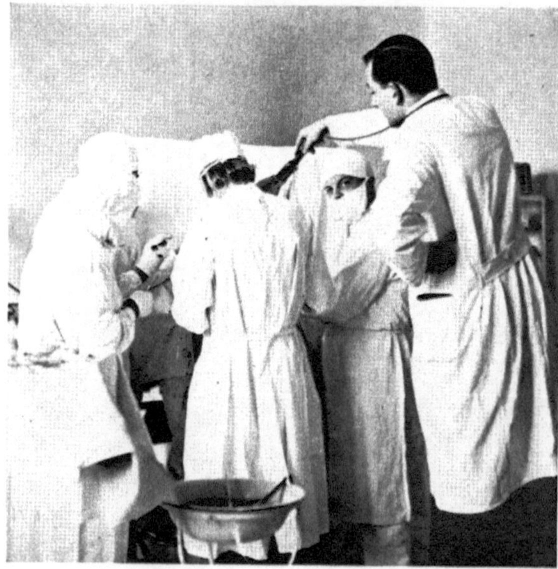

Fig. 10. "He kept on working without silver clips, without electric drill, without suction or electro-cautery, and sometimes employing an awkward and disadvantageous positioning of the patient on an old operating table"

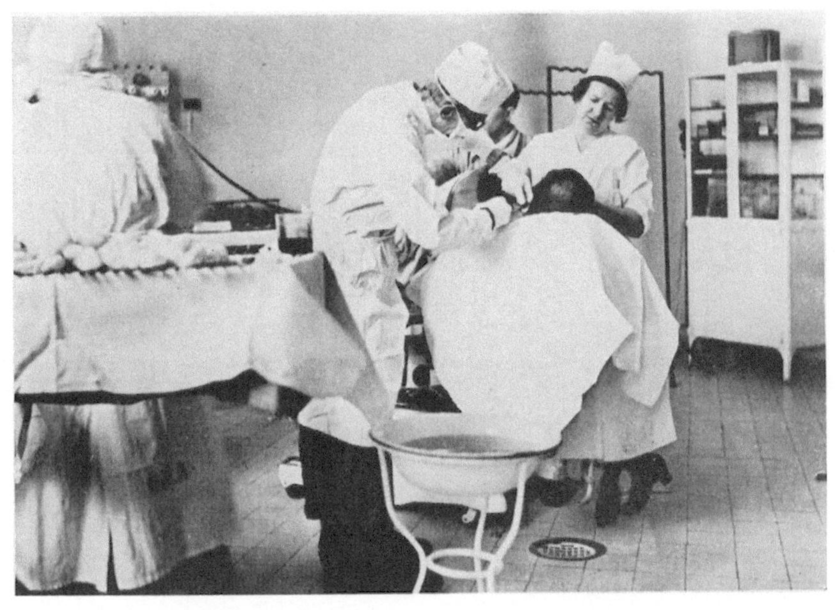

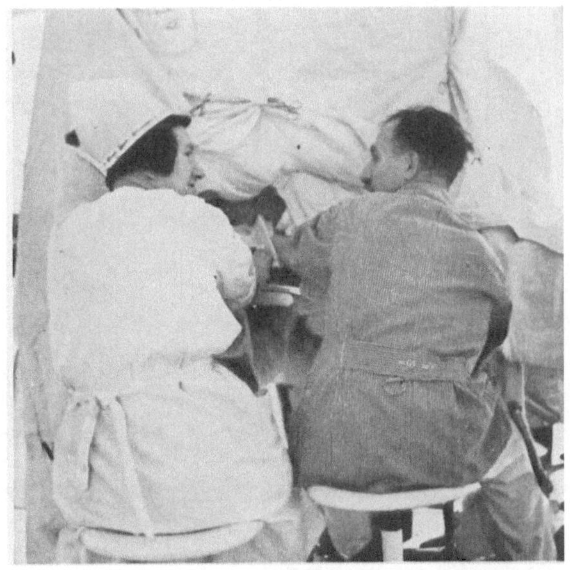

Fig. 10

Breslau
Hohenlohestr 11
18 V 35

Dear Doctor Kahn,

Many thanks for your letter of May 8th. I have performed the splanchnic nerve sections June 1924. I resected in addition to ~~both~~ the splanchnic nerve the 6–10th ganglia of the sympathetic chain. I exposed the thoracic sympathetic chain and the splanchnic nerve ~~after~~ by removing the transversal processes of the corresponding vertebrae and the capitula of the ribs. The pleura was carefully pushed aside. I performed the operation on both sides. Note that I resected but both ~~major~~ major splanchnic nerves, not the minor splanchnics. (11. 12. R.).

I am sorry to say that I can not give you the reference of Jean's paper. It is so long ago that I wrote that book on pain and I do not remember in which journal Jeans paper is published.

Cordial greetings to Dr Peet and to yourself.

Yours always
sincerely
Otfrid Foerster

Fig. 11. Description of the first bilateral resection of the major splanchnic nerves and of the 6th to 10th thoracic ganglia of the sympathetic chain, in a letter to Dr. E. KAHN (Ann Arbor, Michigan)

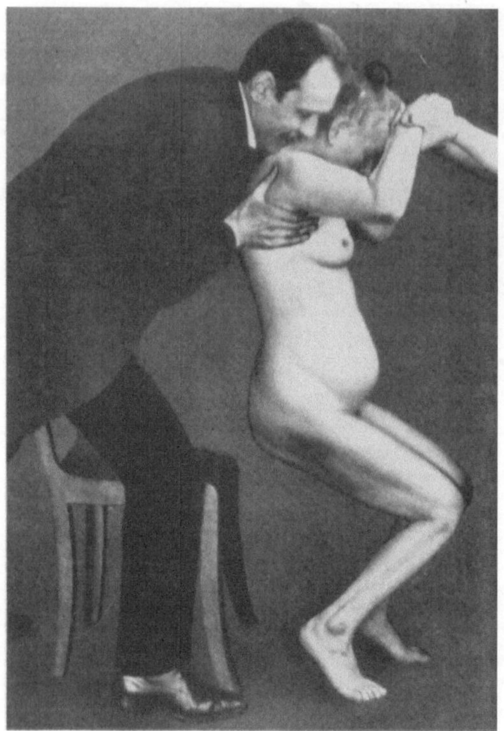

Fig. 12. "FOERSTER became the master photographer of neurology. Numerous early pictures have been preserved in which he holds the patient and guides her as an artist holds his cello. Never thereafter do we read in his finely outlined but still youthful face his zeal and his unbroken joy in life . . ."

Fig. 13. "... In order to keep his memory alive, the Otfrid Foerster Medal and a lectureship were established by the German Society of Neurological Surgery"
(A copy from the original of the Foerster Medallion, which was destroyed)

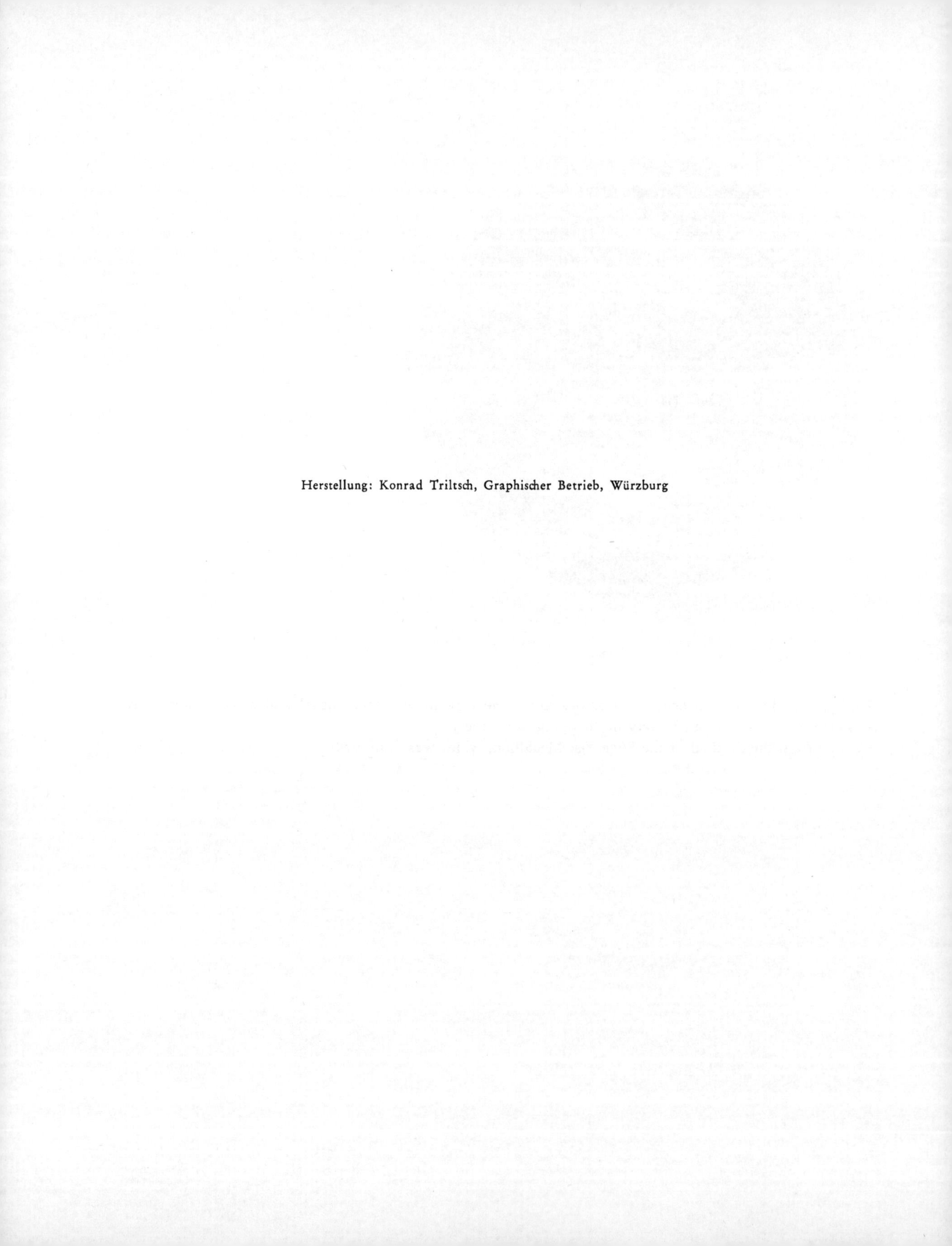

Herstellung: Konrad Triltsch, Graphischer Betrieb, Würzburg